AF717555

Thieme

# Introduction to Bioregulatory Medicine

Alta Smit, MD
General Manager
International Academy of Homotoxicology (IAH)
Baden-Baden, Germany

Arturo O'Byrne, MD
Head of Clinic
Centro de Medicina Biológica
Cali, Colombia

Bruno Van Brandt
Medical Education Manager
International Academy of Homotoxicology (IAH)
Baden-Baden, Germany

Ivo Bianchi, MD
President
Italian Society of Homotoxicology
Verona, Italy

Klaus Kuestermann, MD
President
International Association for Biomedicine
Baden-Baden, Germany

80 illustrations

Thieme
Stuttgart · New York

*Library of Congress Cataloging-in-Publication Data*

Introduction to bioregulatory medicine / Alta Smit ... [et al.].
p. ; cm.
ISBN 978-3-13-147611-1
1. Homeostasis. 2. Physiology, Pathological. 3. Extracellular matrix. I. Smit, Alta.
[DNLM: 1. Homeopathy–methods. 2. Homeostasis.
WB 930 I616 2009]
QP90.4.I58 2009
612'.022–dc22
2009017692

**Important note:** Medicine is an ever-changing science undergoing continual development. Research and clinical experience are continually expanding our knowledge, in particular our knowledge of proper treatment and drug therapy. Insofar as this book mentions any dosage or application, readers may rest assured that the authors, editors, and publishers have made every effort to ensure that such references are in accordance with **the state of knowledge at the time of production of the book.**

Nevertheless, this does not involve, imply, or express any guarantee or responsibility on the part of the publishers in respect to any dosage instructions and forms of applications stated in the book. **Every user is requested to examine carefully** the manufacturers' leaflets accompanying each drug and to check, if necessary in consultation with a physician or specialist, whether the dosage schedules mentioned therein or the contraindications stated by the manufacturers differ from the statements made in the present book. Such examination is particularly important with drugs that are either rarely used or have been newly released on the market. Every dosage schedule or every form of application used is entirely at the user's own risk and responsibility. The authors and publishers request every user to report to the publishers any discrepancies or inaccuracies noticed. If errors in this work are found after publication, errata will be posted at www.thieme.com on the product description page.

© 2009 Georg Thieme Verlag,
Rüdigerstrasse 14, 70469 Stuttgart, Germany
http://www.thieme.de
Thieme New York, 333 Seventh Avenue,
New York, NY 10001, USA
http://www.thieme.com

Cover design: Thieme Publishing Group
Typesetting by Sommer Druck, Feuchtwangen, Germany
Printed in India by Replika Press Pvt. Ltd.
ISBN 978-3-13-147611-1

2 3 4 5 6

MIX
Paper from responsible sources
FSC www.fsc.org FSC® C016779

Some of the product names, patents, and registered designs referred to in this book are in fact registered trademarks or proprietary names even though specific reference to this fact is not always made in the text. Therefore, the appearance of a name without designation as proprietary is not to be construed as a representation by the publisher that it is in the public domain.

This book, including all parts thereof, is legally protected by copyright. Any use, exploitation, or commercialization outside the narrow limits set by copyright legislation, without the publisher's consent, is illegal and liable to prosecution. This applies in particular to photostat reproduction, copying, mimeographing, preparation of microfilms, and electronic data processing and storage.

# List of Abbreviations

| | |
|---|---|
| ADP | adenosine diphosphate |
| ACTH | adrenocorticotropic hormone |
| APC | antigen-presenting cell |
| ARS | auto-regulating system |
| ASA | acetylsalicylic acid |
| ATP | adenosine triphosphate |
| BALT | bronchial-associated lymphoid tissue |
| BBRS | basic bio regulating system |
| CD | cluster designation |
| COX | cyclooxygenase |
| CPG | catalyst preparation groups |
| CRH | corticotropin-releasing hormone |
| DDT | dichlorodiphenyltrichloroethane |
| DET | disease evolution table |
| ECM | extracellular matrix |
| FAD | flavin–adenine dinucleotide |
| $FADH_2$ | flavin–adenine dinucleotide reduced form |
| FMO | flavin-containing monooxidases |
| GABA | γ-aminobutyric acid |
| GAG | glycosaminoglycan |
| GALT | gut-associated lymphatic tissue |
| GAS | general adaptation syndrome |
| GIT | gastrointestinal tract |
| GRS | ground regulation system |
| GSH | reduced glutathione |
| HLA | human leukocyte antigen |
| HOT | hematogenic oxygenation therapy |
| IM | intramuscular(ly) |
| IRD | inflammation-regulating drug |
| IV | intravenous(ly) |
| LGL | large granular lymphocyte |
| MALT | mucosal-associated lymphatic tissue |
| NAD | nicotinamide–adenine dinucleotide |
| NADH | nicotinamide–adenine dinucleotide reduced form |
| MHC | major histocompatibility complex |
| NOAL | no observed adverse effect level |
| NSAID | nonsteroidal anti-inflammatory drug |
| ORG | organ preparation groups (organ regulator) |
| PBBE | polybrominated biphenyl ether |
| PCB | polychlorinated biphenyl |
| PG | proteoglycan |
| Pi | inorganic phosphate |
| PNEI | psycho-neuro-endocrine immunology |
| PPG | plant preparation group |
| RCD | regulation/compensation division |
| RES | reticuloendothelial system |
| ROS | reactive oxygen species |
| SBS | sick building syndrome |
| SC | subcutaneous(ly) |
| TCR | T-cell receptor |
| TGF | transforming growth factor |
| $T_H1$, $T_H2$ | T-helper subsets |
| TNF | tumor necrosis factor |
| $T_{reg}$ | T-regulatory cells |
| UV | ultraviolet |

# Preface

Hans-Heinrich Reckeweg, a German physician, first conceived the theory of homotoxicology more than five decades ago. During the succeeding 50 years, many of the elements of a scientific basis for bioregulatory medicine were discovered and described. Alfred Pischinger performed basic holistic research and described the ground system in the 1970s. This was followed by the histological research of Hartmut Heine on the basic structures of the extracellular space, a histological fact that is fully accepted in modern science. Five decades of implementation of Reckeweg's original concept of insights into disease and disease evolution have passed, leaving us today with the valuable concept of homotoxicology. Finally, and especially in the last few years, basic research has been undertaken, confirming hypotheses that had formerly been expressed for years, even decades, by experts in homotoxicology worldwide.

Most of this advanced basic research elucidates how microconcentrations and nanoconcentrations of components in antihomotoxic medications can influence steering processes in the body. The correcting and steering influence on mediator-secreting immune cells has been studied in depth. Cell protective effects have been shown by the use of minute concentrations of components used in antihomotoxic medications.

In more than 50 years, the homotoxicological approach to disease and antihomotoxic treatment has gained tremendous ground, not only in general practice but also in specialized medical branches. This is mainly because of its high accessibility for conventional medical doctors and for complementary medicine-oriented practitioners. Logically, and not in contradiction to the basic principles of medicine, homotoxicology has made its way into many practices and therapeutic applications.

Science is dynamic, constantly evolving, and the same holds true for homotoxicology. Accordingly, this book does not pretend to be complete; rather, its aim is to be a reference, a study book in homotoxicology. It is an up-to-date resource on what we know about homotoxicology today: what is evidence based and what hypotheses and issues are still open to be analyzed and solved.

We invite you to enter, observe, and study the world of homotoxicology with the same open mind as a scientist would have in observing and examining a new phenomenon. It may be that you encounter contrasts with what you have studied until now, that you will discover new dimensions to what you thought was a fixed and closed issue.

This book may provide a different viewpoint on what disease and health are and on what evolution of the patient is or is not healthy. These considerations may influence the treatments you use and may help you approach the patient more holistically. The book may lead you to be a better practitioner, one who has full respect for and harmony with the basic principles of life. Your patients will feel the difference and be grateful.

Homotoxicology is not a dogmatic concept and is by no means an absolute in any sense of the word. It is open for integration at all levels—a supplementary reason why it is increasingly accepted in modern medicine. Homotoxicology will add an extra dimension to your profession as a health care practitioner. We hope that you, along with thousands of other practitioners worldwide, will look back in a few years and be grateful that you learned about this new theory.

*Medicine is the only profession that continuously fights against the reasons for its own existence.*

James Bryce, 1838–1922.
British statesman, jurist, and historian

*Alta Smit*
*Arturo O'Byrne*
*Bruno Van Brandt*
*Ivo Bianchi*
*Klaus Kuestermann*

# Contents

# 1 Introduction to Bioregulatory Medicine

The human body is an extremely complex, multiplex interaction system of 100 000 billion cells. The human brain has an average of 100 billion cells, each of them interacting with thousands of other cells. That is the potential for 100 000 billion interactions, a number equal amount to the sum of all the stars in 1000 galaxies. The Purkinje cell further illustrates this complexity. The 150 000 to 175 000 cell inputs received by each Purkinje cell[1] constitute the most massive synaptic convergence found on any neuron in the brain[2]. In addition to the steering activities of the nervous system, other systems intervene in bodily processes. For example, the endocrine system releases hormones and the immune defense system uses steering cytokines and other mediators to organize and steer defense. The most remarkable fact is that all these bodily systems work together as a synchronized unit to maintain a steady state. It is amazing, to say the least, that these billions of interactions per second do not lead to complete chaos in the body. Instead, they lead to a smoothly running, dynamic system that can work for decades.

Life, indeed, is a dynamic given. Although momentary observations (e.g., blood glucose level, T-cell count, and hormone level) may suggest a static state, life is above all characterized by continuous changes in these bodily processes, making it dynamic.

To stay as close as possible to the main objective, the steady state of the organism, most life processes within the human body are steered by interactive autoregulating systems (ARSs), often using feedback systems to steer and correct their outcomes. Therapeutic interventions by which these processes are influenced, inhibited, or stimulated (with full respect for the biological steering character of the systems themselves) are part of bioregulatory medicine.

> ***Bioregulation** means regulation of biological processes (Greek **bios** = life).*

On first sight, the human organism may look like a closed microenvironment, but humans are continuously influenced by their surrounding macroenvironment. Interactions between the microenvironment and the macroenvironment will disturb the stable inner environment, triggering correcting actions of the steering autoregulating processes of the body, seeking a steady state again.

To understand the basics of bioregulatory medicine, we have to first study the basic principles of bodily regulation toward a steady state, called *homeostasis*. The main objective of this book is to understand the basic principles of bioregulatory medicine and an integral part of it, homotoxicology. Terms such as *autoregulating systems, homeostasis, internal environment, feedback system*, and *open and closed systems* are all key in bioregulatory medicine, which is why these terms need to be explained in more depth and detail.

---

[1] Harvey RJ et al. Quantitative studies of the mammalian cerebellum. *Prog Neurobiol.* 1991;36:437–463

[2] Shepherd GM. *The Synaptic Organization of the Brain.* New York: Oxford University Press; 1990

## Autoregulating Systems

The human body is made of extraordinarily unstable material, sensitive to the smallest stimuli from outside. To live is to be continuously in interaction with the surrounding environment. Changes in this direct environment of the body have an immediate effect on the internal milieu. On the other hand, very subtle stimuli are present and active in the body at all times, amplifying or inhibiting the external triggers. In both ways, body and environment form one extremely complex interacting unit, changeable all the time.

Stimuli so minute that very delicate methods—such as electrocardiography, electroencephalography, or myography—are required to measure them course along our nerves. At the end of the nerve, a muscle is so sensitive to this subtle stimulus that it reacts with a powerful contraction, resulting in a firm movement of a body part in its external environment. On first sight, the force of this movement is not in relationship to the dosage of the stimulus: a small electrical impulse as a result of minute dosages of steering neurotransmitters. Our sense organs are responsive to almost incredibly minute stimulations. Microconcentrations and nanoconcentrations of an odoriferous compound are detected in the air and can trigger our attention, even alert us to danger in the environment. The same is the case with subtle noises, perhaps even more so: a complete absence of noise may trigger an alarm state in most mammals in nature. Similarly, the minute pressing of one single hair on our face can trigger itching sensations and attract our full attention for a moment.

All of these examples have one aspect in common: the triggering impulse is subtle, often caused by minute doses of a substance, and the reaction of the organism to the trigger is managed and directed through minute dosages of steering mediators.

Although always seeking harmony, the balance of the human organism is constantly disturbed by these subtle or sometimes more obvious stimuli, present in all gradations possible. The organism is continuously trying to correct these deviations to regain its steady and harmonious state. To be able to do so, the organism possesses autoregulating systems that can "direct," "correct," or "manage" most bodily processes, using subtle quantities of mediators or impulses to do so. Most of the bodily processes steered by these autoregulating systems are not isolated at all; they are in continuous interaction with each other, forming a uniform and harmonious reaction against environmental triggers and disturbances.

What is remarkable is that even if the disturbing stimulus is huge, the autoregulating systems continue to use subtle, minute dosages of steering mediators to correct deviations. Although these autoregulating systems might appear as relatively autonomous and are often presented as such in the literature, most bodily processes interact with each other, often reciprocally, and the synergy of the systems all together seeks to achieve the main common objective: to restore an organism-entity to steady state. It is the subtle balance of health on the thin cord of life.

### Definition of an Autoregulating System

An autoregulating systems (ARS) is a complex, cybernetic system that merges different disturbing and steering stimuli. Information loops are often used, whereby the output of a subsystem will influence the input in such a way that the outcome becomes iteratively closer to the variable set point of the system.

The ultimate and only objective of the ARS is to keep the outcome of the process as close as possible to the variable set point. In this way, we may state that *health* can be defined as a state in which all the bodily processes, at the same time, are within their variable set point margins. *Disease*, in the same way, can be defined as a dysregulation, a deviation of one or more processes and their ARSs from the variable set point. In disease, the steady state is (temporarily) lost. Bioregulatory medicine will intervene to correct the deviations in such a way that the steady state is restored.

The important features of bioregulatory medicine are the variable set point, the changing environment influencing the steady state (perturbing the outcome of the system from its set point), the correcting physiological reactions of the organism to restore the steady state, and the final state of homeostasis of the whole organism in its environment.

### Set Point Objective and Fluctuating Reality

Most physiological processes have a variable set point or a narrow or even a broad set point range, adaptable to counter environmental changes and stimuli. Although the set point can be an optimized value in some cases, in most physiological processes there is a variable set point band that tolerates some fluctuation of the values between a minimum and a maximum value before a correcting intervention is triggered. Metaphorically, a set point of 21 °C on a thermostat will probably trigger the heating system to heat at 20.8 °C and will signal the system to stop heating at 21.2 °C. Although 21 ° C is the set point, the system will show a "tolerance zone" of deviation. The more precise and critical the system (and a biological process is a system), the smaller the set point margin.

Optimized and at rest, humans will have a respiration rate of about 12–15 respirations a minute, which immediately increases upon a detected increased carbon dioxide ($CO_2$) level in the blood. Heart rhythm in adults at rest will be between 60 and 100 beats a minute. Trained athletes may have a much lower set point: about 45 beats a minute. Movement will immediately result in a changed set point, increasing heart rhythm to counter the changes induced by this movement.

The macroenvironment around humans is constantly changing. Environmental temperature, air pressure, humidity, pollution, and concentration of microorganisms are changing every minute of the day. In this changing environment, humans try to maintain a steady state, keeping physiological processes within the set point margin. In reality, this means that humans will have to adapt temporarily to changed values, correct deviations, and restore conditions.

This fluctuating reality is in strong contrast to the body's "desire" for stable physiological values. Without ARSs in an open system, humans, along with many other biological entities, would not be able to survive and maintain the steady state of health.

### Homeostasis

The concept of multiple dynamic equilibrium adjustments and regulation mechanisms was created by the Frenchman Claude Bernard (1813–1878). Bernard is often considered the father of physiology[3]. "Multiple dynamic equilibrium" is the basic principle behind homeostasis. The term *homeostasis* was defined in 1932 by Walter Bradford Cannon, who derived it from the Greek word *homoios* (which means "the same," "like," or "resembling") and a second word *stasis* (which means stance or posture). Homeostasis literally means "to remain in the same condition," to remain as close as possible to a steady state of the system.

Bernard described the internal environment of the human body as directly influencing the state of the living cell[4]. Carl Rokitansky, a Czech medical scientist (1804–1878), was considered the father of objective experimental pathology. He believed the blood was the cause of diseases. He divided pathological conditions into *crases* (deficits) and *stases* (depositions or accumulations) and related diseases and organ degeneration to the humoral cell's environment. Another scientist of that time was the German Rudolf Virchow (1821–1902), who pointed to the cell as the smallest living unit of the organism. Virchow founded the medical discipline of cellular pathology and stressed the importance of cell function in diseases.

Bernard had a more integrative mind and stressed the importance of transport of essential substances to and from the cell to keep its function and condition optimized. He understood the importance of steady conditions in the direct environment of the cell. In this way, Bernard was countering Virchow, who stressed the cell as the smallest living unit in the organism and its function as essential for determining between health and disease. Virchow related diseases directly to

---

3 Bernard C. *An Introduction to the Study of Experimental Medicine*. 1865. [First English translation by Henry Copley Greene, published by Macmillan & Co., Ltd., 1927; reprinted in 1949.]

4 Bernard C. *Leçons sur les Propriétés Physiologiques et les Altérations Pathologiques des Liquides de l'Organisme*. Vol. 2. Paris: Bailliére; 1859

the cell[5]. Unlike Rokitansky though, he did not stress the importance of the cell's environment for the cell's health. Bernard related the one to the other, stressing the importance of the quality of the cell's environment for its effective functioning.

As mentioned above, the concept of homeostasis, or steady state of outcome of bodily processes, was introduced much later (in 1932) by Cannon, an American physiologist[6]. Homeostasis has become one of the central tenets of every theory of self-regulation and internal organization in complex open systems.

*Cannon defined homeostasis as "the coordinated physiological reactions which maintain most of the steady states in the body."*

In this definition, we recognize the dynamic processes, via physiological reactions to external stimuli, for keeping in the body a steady, stable state of all the physiological processes present.

The word homeostasis is often used in medical literature to denote a healthy condition and balance of all bodily processes.

## Cybernetics: Negative, Positive, and Bipolar Feedback Systems

Cybernetics is the interdisciplinary study of the structure of complex systems, their steering mechanisms (especially communication processes and information flows), control mechanisms, and steering corrective feedback principles. Cybernetics is commonly related to steering mechanisms in machines and robots, automation processes used in industry, but it is also the study of the steering processes in living organisms, called *biological cybernetics* or *biocybernetics*. Being a complex open system in continuous interaction with its environment, the human body is an excellent study object in biological cybernetics. Another subdiscipline of cybernetics, related to human beings, is *psychocybernetics*, the study of the steering and interacting factors and feedback loops behind mental processes and human behavior.

Within cybernetics, the following subcategories are recognized:

- Biocybernetics is the study of an entire living organism.
- Neurocybernetics is the study of neurological models and neurotransmitter interactions.
- Molecular cybernetics is the study of molecular systems (e.g., molecular biology cybernetics).
- Cellular cybernetics is the study of cellular systems (e.g., information technology/cell phones or biological cells).
- Evolutionary cybernetics is the study of the evolution of informational systems.

Generally three types of feedback systems are discussed in cybernetics: negative, positive, and bipolar feedback systems. Negative and positive feedback systems were first introduced by Norbert Wiener[7]. Bipolar feedback systems are a merging of positive and negative feedback in one system, using two or more parallel loops simultaneously or in continuous alternation.

Negative and positive feedback loops have a kind of on/off switch to induce correcting action and then to stop this action if the set point is reached or the outcome value of the whole system is between acceptable margins, as illustrated in **Fig. 1.1**.

*Negative Feedback*

In negative feedback, the output of the system will fully or partially influence the input in such a way that it tends to reduce the outcome of the system. Pure isolated autonomic negative feedback systems do not occur in the healthy human body, only as correcting subsystems in more complex ARSs. Cortisol release can be used as an example of a negative feedback loop in an autoregulating bodily process. In the hypothalamus–hypophysis–suprarenal cortex axis, a certain level of cortisol concentration reached in the blood

[5] Virchow R. *Die Cellularpathologie*. Berlin: Verlag von August Hirschwald; 1858

[6] Cannon WB. *The Wisdom of the Body*. New York: Norton; 1932

[7] Wiener N. *Cybernetics or Control and Communication in the Animal and the Machine*. Paris: Hermann et Cie–Cambridge, MA: MIT Press; 1948

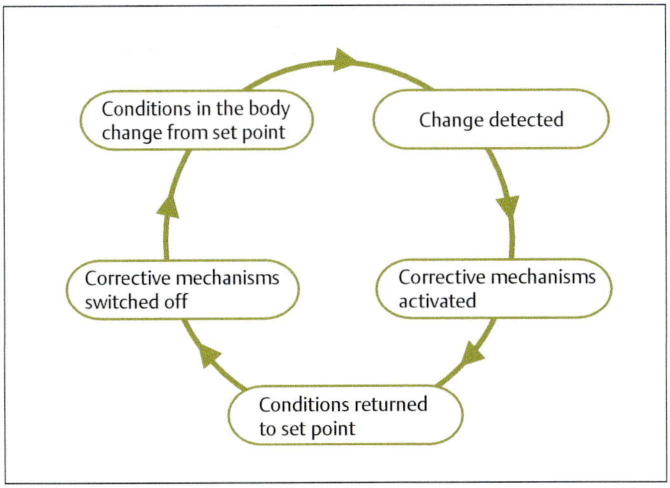


**Fig. 1.1** Negative and positive feedback loops.

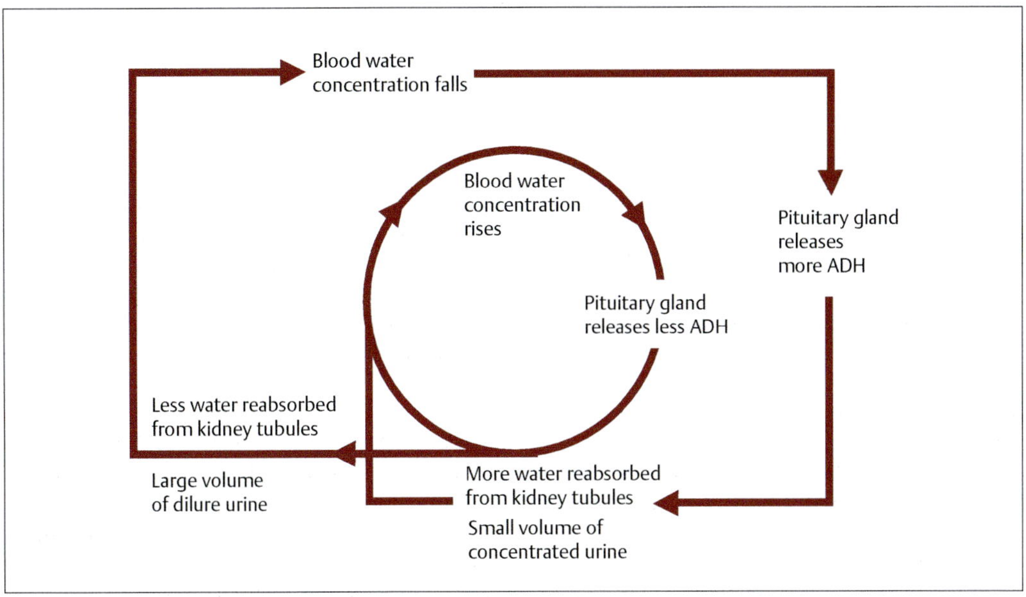


**Fig. 1.2** Osmoregulation: the control of water concentration in blood and bodily fluids.

(outcome) will influence the input of the system in such a way that less cortisol is released by the suprarenal glands.

Another example of negative feedback can be seen in *osmoregulation* (the control of water concentration in blood and bodily fluids). **Figure 1.2** demonstrates the feedback loop used to control and regulate water fluid concentration in kidney activity.

*Positive Feedback*

In positive feedback, the output of the system will fully or partially influence the input in such a way that it tends to increase the outcome of the system. The end result of a positive feedback system is often explosive because each step or full loop of the process will amplify the former step. When not in combination with an inhibiting negative feedback system (as is the case in a bipolar feed-

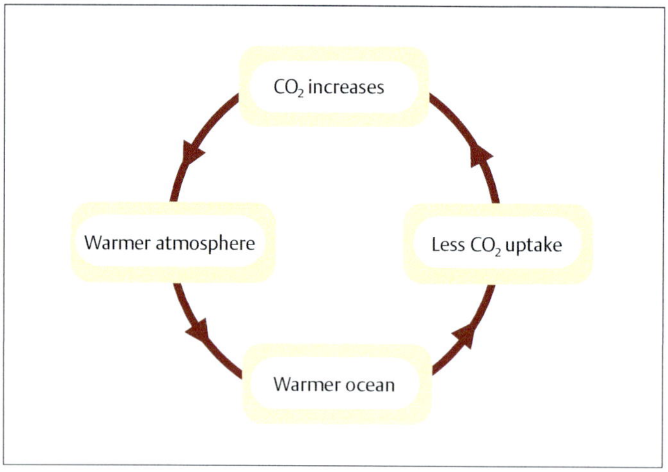


**Fig. 1.3** A positive feedback loop: $CO_2$ emission causes heating of the earth.

back system), the isolated positive feedback system will often be self-destructive. For example, bombarding fissile material with neutrons causes it to emit even more neutrons, which, in turn, affects the material even more, causing the emission of yet more neutrons. Every time the system goes through the loop again, the outcome value increases.

Another example comes from acoustics in what is what is known as the Larsen effect after the Dane Soren Larsen, who first discovered and described the principle of audio feedback. The sound emitted from a loudspeaker is fed back to the microphone and is amplified again and again, causing an extremely loud whistling effect. The loudspeaker can ultimately be destroyed if no outcome-limiting system, such as peak-clipping, is foreseen and implemented in the electronic circuit. The Larsen effect as a positive feedback system is one of the main problems in audiology: deaf persons may use an all-in-the ear prosthesis in which microphone and speaker are very close to each other (sometimes < 1 cm apart), which can cause annoying whistling sounds for the patient.

A final example is the positive feedback loop that causes heating of the earth by $CO_2$ emission, as illustrated in **Fig. 1.3**. Meteorologists now predict that this positive feedback loop will seriously endanger future life on earth because of this warming phenomenon.

Positive feedback systems do not occur in healthy conditions in human beings because they would lead to excessive, life-endangering outcomes. They may play a role as the bases of some pathological processes (e.g., fear-induced tachycardia causing fear itself, and this fear triggering an increased heart rate as a consequence).

Generally, we may say that the negative feedback loop tends to slow down a process, whereas the positive feedback loop tends to speed it up, often to explosive, self-destructive results. In bipolar feedback loops, positive and negative feedback systems may occur together, inducing reciprocal inhibiting effects controlling the system.

*Bipolar Feedback*

In bipolar feedback, the output of the system will increase or inhibit the input in such a way as to produce increased or decreased functional outcome in relation to the needs of the organism. Bipolar feedback systems occur in the human body, often interacting with each other (reciprocal inhibition or alternating action). Physiological processes, with their outcomes, may influence other processes, which are again reciprocally influenced by them. If both (or more) interacting processes have a bipolar feedback system, subtle changes in one will induce subtle changes in the other (and vice versa).

### Open and Closed Systems

Both positive and negative feedback are seen as closed systems in cybernetics. They are called *closed systems* because the system is closed by a feedback loop (i.e., the response of the system depends on the feedback signal alone to complete its function). Without such an isolated loop, it would become an open system because external stimuli might interfere and influence the input too. In contrast, a feed-forward system is an open system because it does not have any isolated feedback loop and does not rely on a feedback signal alone to perform its function. In a feed-forward system, the one condition or stimulus will trigger the process to step forward in its next step in the whole process (e.g., catalysts in the citric acid cycle). Catalysts are, thus, enhancing factors to feed forward the Krebs cycle. Feed-forward systems are mostly extremely complex because many conditions need to be fulfilled at the same time to begin the process.

Closed systems are not influenced by external changes or stimuli. Even the cortisol loop, described as a negative feedback system and often defined as a closed system, is not closed at all in humans because other environmental stimuli might influence the final level of cortisol (e.g., stress and inflammatory processes).

The human body is defined as an open system. Internal and external stimuli can influence the steady state of the organism, and ARSs will react by correcting and steering the process to move the outcome once again near to the set point margin.

### Wave Pattern within the Set Point Margins

Variable values, due to influences of all kind, corrected by steering processes in the body will show an oscillating wave pattern (**Fig. 1.4**). This oscillation will be within the set point limits in a healthy condition and will move out of the tolerance band in pathological processes, showing characteristic reactions of the organism, often exhibited as clinical symptoms.

The cause for the wave pattern is the slowness of the reaction of the ARS each time it passes through its set point. Once disturbed by internal or external triggers, it takes time for the system to detect and react accordingly when the process runs below or above the set point or gets outside of the set point margin. The amplitude of oscilla-

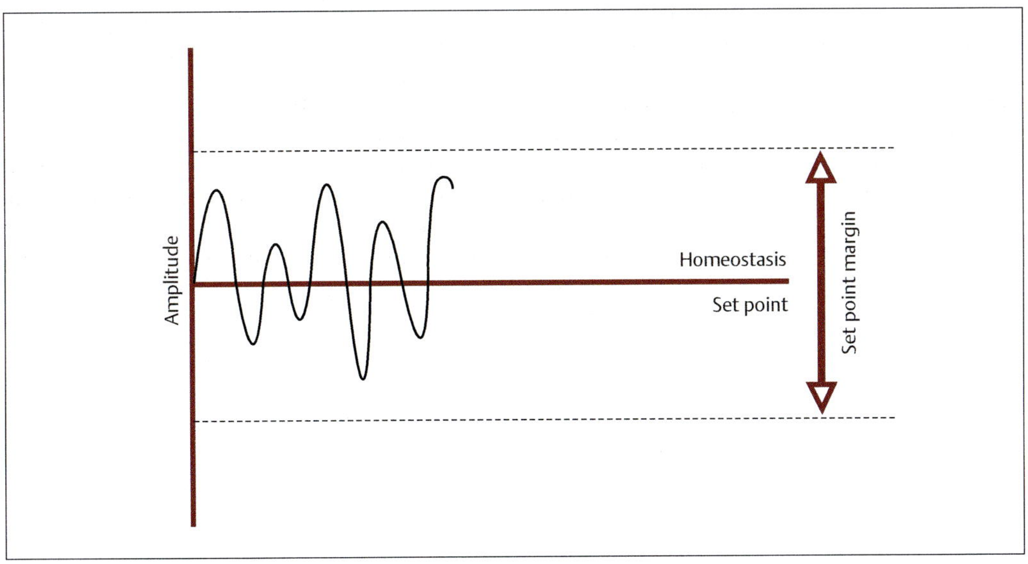


**Fig. 1.4** An oscillating wave pattern within the set point margin.

tion around the set point depends on the accuracy or sensitivity of reaction of the system itself. Coarse systems will have large wave patterns showing high-amplitude oscillations around the set point, whereas subtle and fine-tuned systems will have small wave patterns and accordingly small fluctuations.

The more intense the external disturbing stimuli present, the bigger will be the oscillations around the set point at the onset. Each oscillation tries to correct the deviation back to the set point value. The process can be compared with the behavior of a car on slippery ice. The actions of the driver of a car slipping on ice, as observed from above, will show a wave pattern. Deviation to the right will trigger a leftward steering reaction of the driver, making the car cross its set point, which in this case would be the path straight ahead. Deviation to the left, after crossing the set point, will then be corrected by the driver by steering right, crossing the set point again, this time a little bit less than the first time. This correction will go on and on until the set point is reached, which is driving straight ahead with the car on ice once more.

Along with any other cybernetic system, any bodily autoregulating process can become rigid, which means it can lose its ability to counteract or adapt, resulting in a state of being stuck in a position deviating from the set point. This phenomenon will express itself as abnormal and, depending on the reason for the existence of this process, will show corresponding anomalies. Although this stuck state can occur at any stage of the process, in the human body often top or base rigidity will appear, meaning that the system becomes stuck at the top of the wave pattern or at the bottom of it, respectively.

This happens because many interactions exist between bodily ARSs, and if one system is dysfunctional, the balance will be fully in favor of the others. If two opposite systems in a bipolar system—agonist and antagonist—regulate a process, and the agonist becomes dysfunctional, the process becomes stuck in favor of the antagonist, keeping the balance stuck in a deregulated state.

A simplified example might be the subtle balance between $T_H1$ and $T_H2$ T-helper lymphocytes in an inflammatory process. This process is discussed in more detail in Chapter 4, under Immunomodulation; here we just illustrate the balance and its oscillations around the set point. The $T_H1$ lymphocytes release interferon gamma (IFN-γ) to inhibit the $T_H2$ pathway. On the other hand, $T_H2$ cells will secrete an interleukin (IL-10) to inhibit the $T_H1$ pathway. Minute dosages of both mediators (IFN-γ and IL-10) will keep the balance between cellular defense ($T_H1$ pathway) and humoral defense ($T_H2$ pathway) in homeostasis. If the $T_H1$ cells become dysfunctional, less IFN-γ might be released, inducing an excess of $T_H2$ activity as seen in allergies. In reality, the process of keeping the balance between $T_H1$ and $T_H2$ activity is much more complex because many other mediators may intervene in the final model.

In another example, simple movement is induced by fine-tuned tonus regulation in agonist and antagonist muscles. Bipolar deviations will result in a tremor, and unipolar deviations will result in a cramp or the opposite, complete absence of a tonus, both shown as no movement at all.

### *Top and Base Rigidity*

A worst-case scenario in autoregulating processes occurs when the process becomes stuck on its highest or lowest point, creating a top rigidity or a base rigidity, respectively. In principle, the process may become stuck at any possible point in the oscillation, changing from a dynamic pattern into a rigid state, no longer responding to correcting mediators or impulses.

Top rigidity means that down-regulating processes are no longer present or active, or are not accurate enough to counter the up-regulating steering (see **Fig. 1.5**). Also, negative feedback loops are often dysfunctional, not inhibiting stimulating outcomes. This can be a temporary or a continuous dysfunction. Examples of top rigidity might include tachycardia, hyperventilation, muscle cramps, continuous high fever, hay fever, and other forms of allergy.

Top rigidity is seen when an ARS becomes stuck at its upper deviation from the set point, as seen in **Fig. 1.5** when the dynamic wave pattern goes over to a static continuous horizontal line.

Base rigidity is exactly the opposite of top rigidity. The autoregulating process becomes stuck

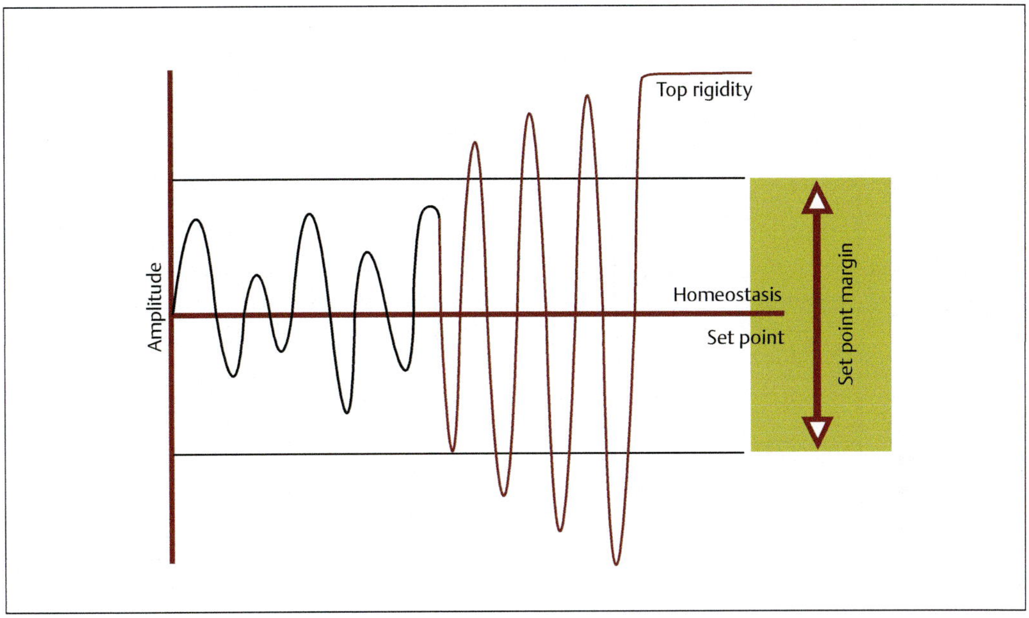


**Fig. 1.5** Top rigidity in an autoregulating process: the process becomes stuck at its highest point.

at its lowest point of the autoregulating activity. In **Fig. 1.6**, the dynamic wave pattern of the process has changed into a static state, represented by the continuous horizontal line at the bottom. Examples of base rigidity include hypoglycemia, depression (serotonin reuptake is too fast), apnea, and cardiac arrest.

## *Examples of Autoregulating Systems*

Most physiological processes in the body are steered through autoregulating processes in an open system. To illustrate this, we briefly discuss several ARSs.

### *Body Temperature*

Many authors[8,9] agree that body temperature has a set point of 37 °C (98.6 °F). Others will define a set point margin[10] or even refer to recent research defining the set point at 36.8 °C. Body temperature may differ by hour of the day, the speed of metabolism of the individual (a fast metabolic rate will result in a higher body temperature, and a slow metabolic rate will result in a lower temperature), and even the body site at which the temperature is measured (oral, anal, or tympanic).

Because all individuals are different, each individual should be viewed as an open system with his or her own ARSs bringing the temperature as near as possible to its variable set point. To maintain the temperature as close as possible to this set point, the organism possesses a complex steering

[8] Campbell NA. *Biology*. 3rd ed. San Fransisco: Benjamin Cummings; 1987:790

[9] McGovern C. "Snatched From an Icy Death". Alberta Report/Western Report, Academic Abstracts, United Western Communications; 1994: 2

[10] Simmers L. *Diversified Health Occupations*. 2nd ed. Canada: Delmar; 1988:150–151

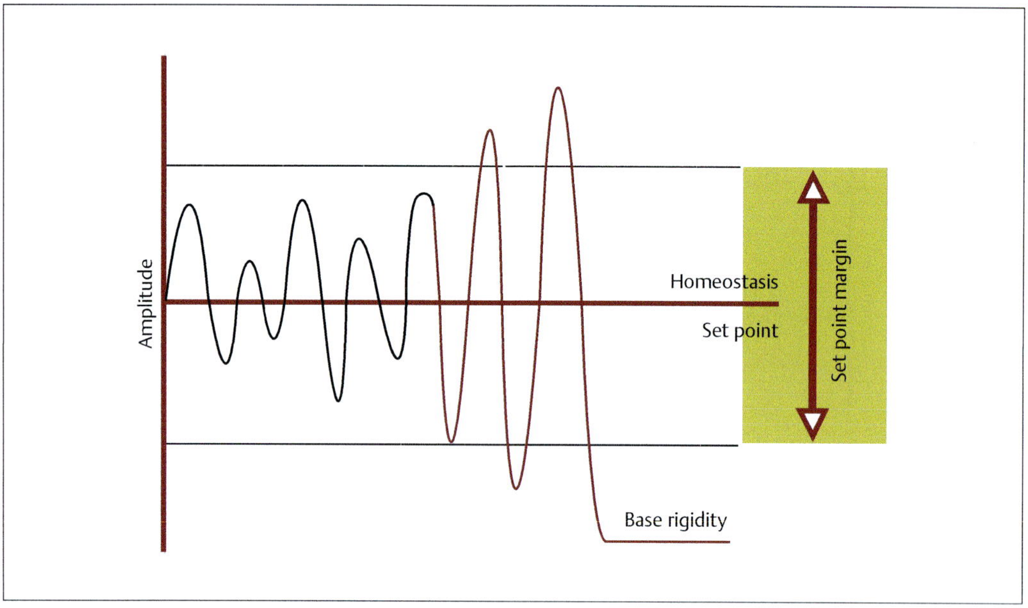


**Fig. 1.6** Base rigidity in an autoregulating process: the process becomes stuck at its lowest point.

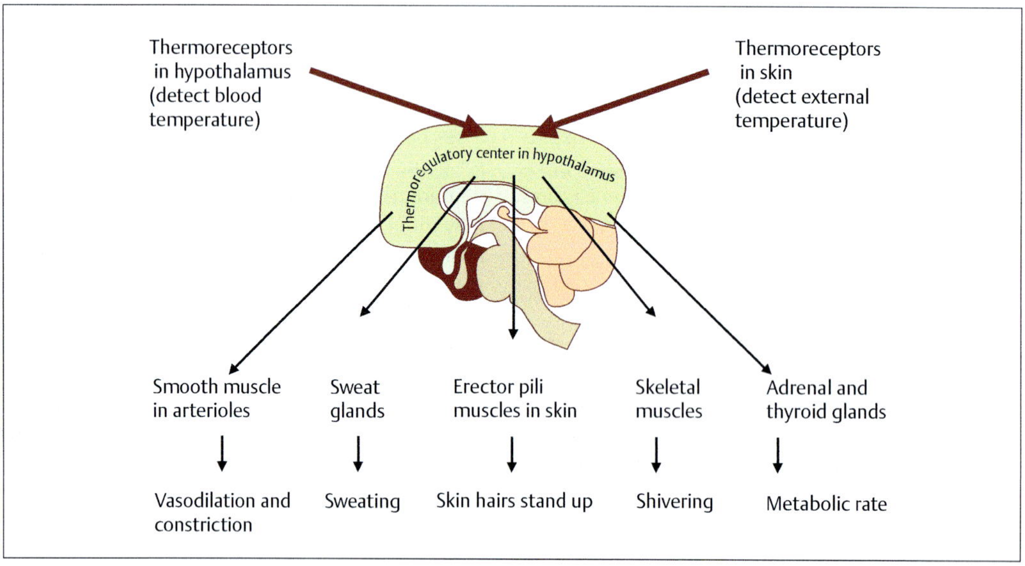


**Fig. 1.7** Thermoregulation.

mechanism that uses sensors to check the process and correct it physiologically if deviations are detected. This phenomenon is called *thermoregulation*. It is achieved via various adaptations and is steered by the hypothalamus, as **Fig. 1.7** illustrates. Thermoregulation is an aspect of homeostasis and is critically important for the proper functioning of the organism.

A synthesis, loops included, is shown in **Fig. 1.8**. Starting from a clear set point of 37 °C at the begin-

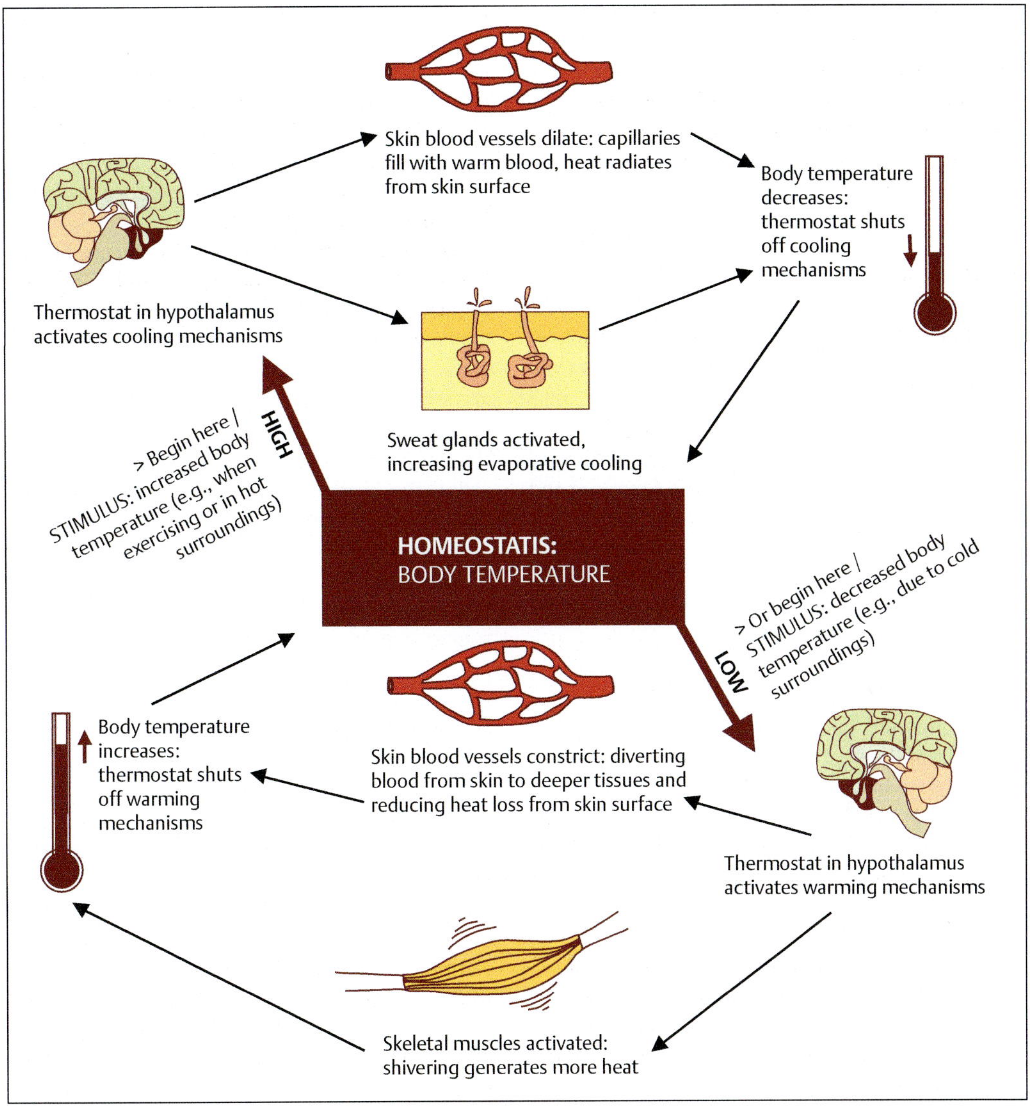


**Fig. 1.8** A synthesis, including loops, of the human thermoregulatory control system.

ning of a feedback loop, the thermoregulating center in the central nervous system (hypothalamus) corrects increased or decreased body temperature. The hypothalamus intervenes in the open system, which is disturbed by external environmental stimuli (warmth or cold in the external environment) or internal steering mediators (cytokines or prostaglandins). Two opposite feedback loops will correct deviations from the set point in a bipolar system.

Various measures are taken to counteract disturbing influences, including

1. Vasoconstriction to decrease the flow of heat to the skin (less body warmth is lost)
2. Cessation of sweating
3. Mechanical heat generation by shivering (which increases the production of heat in the muscles)
4. Secretion of norepinephrine, epinephrine, and thyroxine to increase heat production.

In an open system, such as the human body, the feedback loop of thermoregulation is constantly influenced by external factors. These cause the outcome value to deviate from its set point, triggering correcting mechanisms to bring the outcome back to homeostasis, which lies within the set point margin.

#### *Blood Pressure*

Regulation of blood pressure is a good example of an ARS within an open system. The set point of the system is in most cases 120 mmHg (16 kPa) systolic and 80 mmHg (11 kPa) diastolic. Basically, blood pressure is the result of two forces: the volume of blood forced out of the heart (pump function) and the resistance of the arterial wall. Because this is an open system, all kinds of other internal ARSs may interfere in the final blood pressure outcome. Interacting systems that help keep blood pressure near the set point are as follows:

- Pressure receptors in various organs can detect changes in arterial pressure and adjust the arterial pressure by changing both the force and speed of the heart's contractions, as well as the total peripheral resistance. This is known as the *baroreflector reflex*. The regulatory effect is immediate. Pressure values at the level of the receptors are fed back immediately to correct initial pressure created by the pump (heart) and the heart rhythm.
- The renin–angiotensin system can create a long-term adjustment of arterial pressure. This system allows the kidney to compensate for loss in blood volume or decreases in arterial pressure by activating an endogenous vasoconstrictor known as angiotensin II, which is released to correct the deviation. Internal mediators intervene in the loop.
- A steroid hormone, aldosterone, is released from the adrenal cortex in response to angiotensin II or high serum potassium levels. Aldosterone stimulates sodium retention and potassium excretion by the kidneys. Because sodium is the main ion that determines the amount of fluid in the blood vessels by osmosis, aldosterone will increase fluid retention and, indirectly, arterial pressure. Again, an internal mediator intervenes in the loop.

#### *Blood Glucose Homeostasis*

Glucose is essential for every living cell in the human body. Its concentration is kept strictly between a set point margin of 80 mg/dL and 120 mg/dL (4–8 mmol/L) of blood. Deviations out of this range will cause symptoms such as seen in hypoglycemia and hyperglycemia.

The pancreas is responsible for glucose homeostasis and, as **Fig. 1.9** shows, two opposing loops regulate the concentration. For this purpose, the pancreas has glucose receptor cells (evaluating the actual value and its deviation from the set point) and endocrine cells (islets of Langerhans) that can secrete two oppositely acting hormones: insulin (by the beta cells) and glucagon (by the alpha cells). The hormone insulin has many effects but mainly it stimulates cells to take up glucose and stimulates the liver to convert glucose to glycogen. Insulin directly decreases the concentration of glucose in the blood and the use of fat as an energy source. Its antagonist, glucagon, stimulates the breakdown of glycogen back to glucose, which is released in the blood, directly increasing the glucose levels provided to the cells.

The control and correction processes flow through the two opposing loops, completing a bipolar feedback system. The two loops and their interaction with each other are illustrated in **Fig. 1.9.**

Thermoregulation, blood pressure regulation, and glucose homeostasis constitute only a few examples of all the autoregulating processes acting in the human body. Each of these systems is able, at a certain level, to deal with external influences that disturb homeostasis and to counteract them to restore the steady state in the organism as a whole.

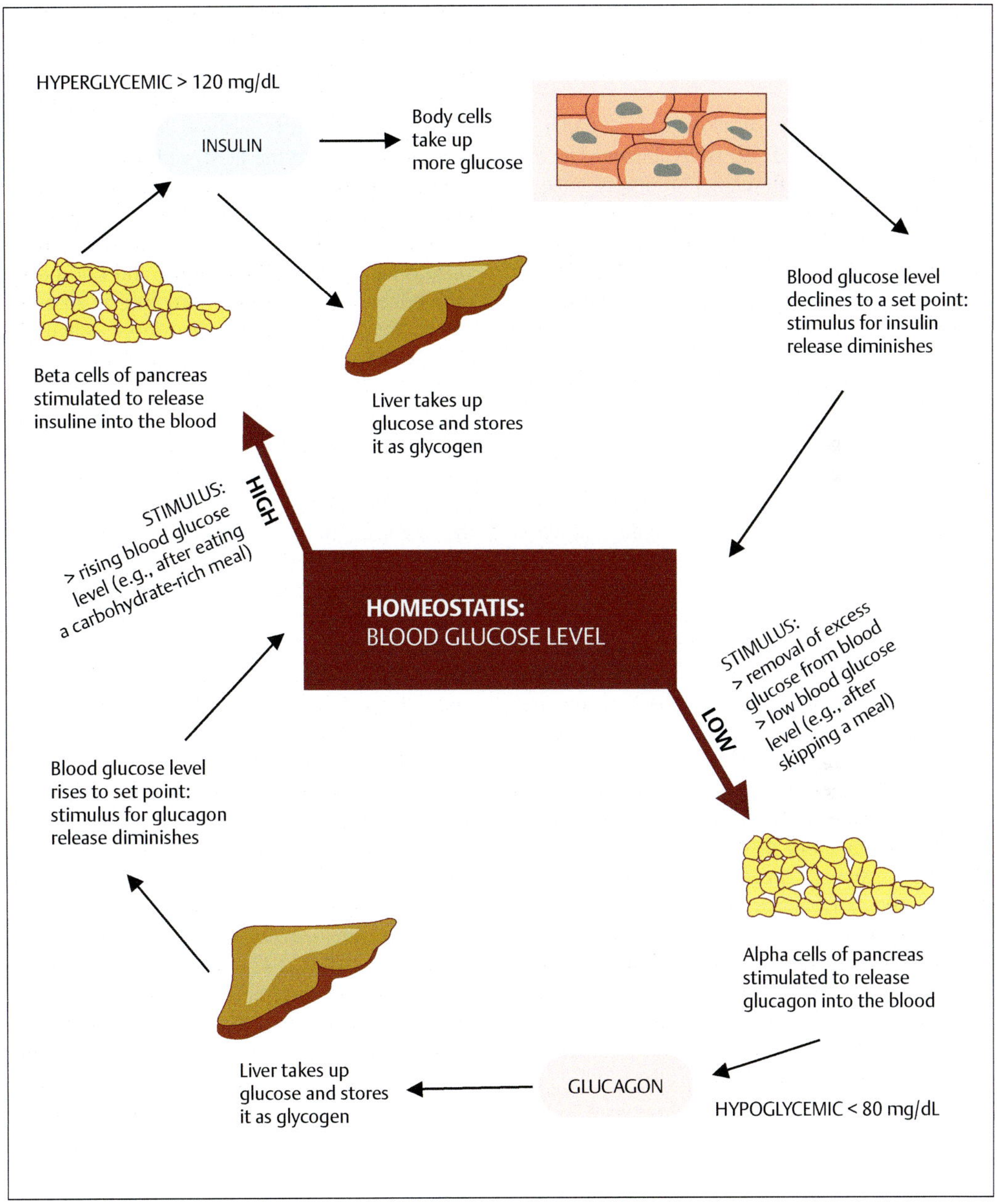


**Fig. 1.9** A synthesis, including loops, of the human glucose homeostasis system.

## The Ground Regulation System of Alfred Pischinger

The histological and physiological research of Alfred Pischinger, an Austrian histologist and embryologist, is of fundamental importance in bioregulatory medicine. In his basic work[11], Pischinger describes the cellular environment and the interactions between different bodily systems that influence each other. In later publications, specifically one edited by Heine in 2004[12], the ground system is characterized as a terrain of exchanges of stimuli (e.g., nerve impulses) and mediators (mainly cytokines and hormones) directing the open system toward a steady state (homeostasis).

What was initially termed the ground regulation system (GRS) by Pischinger is today known as the homeostasis-driven autoregulating processes that mainly occur or have their end effect in the extracellular matrix (ECM).

Although *ECM* is now standard terminology to describe Pischinger's GRS in modern medicine, especially in histology and physiology, other terminology is still found in the literature. The term *basic bioregulating system* (BBRS) is often used[13], especially in central Europe. The term is used within the framework of neural therapy (another type of bioregulatory medicine). This term, introduced by Lamers[14] exactly describes what the ECM is: the extracellular space is "basic" because of its ubiquitous presence in the body. It is "biological" because it generates and maintains the open system in its environment. It is "regulating" because most steering impulses and stimuli will interact via the ECM. Finally, it is a "system" because it conforms to predictable rules.

[11] Pischinger A. *Das System der Grundregulation*. Heidelberg: Haug Verlag; 1983

[12] Pischinger A. *The Extracellular Matrix and Ground Regulation, Basis for a Holistic Biological Medicine*. Heine H, ed. Brussels: Haug Verlag International; 2004

[13] Van Wijk R. *Het Basis Bio-Regulatie Systeem*. Utrecht: Universiteit van Utrecht, Vakgroep Moleculaire Celbiologie; 1987

[14] Lamers H. *Neuraaltherapie en het BasisBioRegulatie-Systeem*. Deventer, Netherlands: Ankh Hermes; 1988

Capillaries and nerve endings do not terminate in the cell itself but rather in the neighborhood of the cell, in the ECM. Cells do not dump their metabolic waste products or mediators directly into the venous or lymphatic system. Rather, these are deposited first into the ECM, and from there they are drained into the venous or lymphatic system. Between the cell and the capillaries and lymph vessels exists a transmission area, filled with fibers such as collagen, elastin, laminin, and fibronectin. In between this main fiber structure is an amorphous, gel-like substance. Pischinger called this substance the *ground substance* (**Fig. 1.10**).

Today, we know that the ECM has many functions. However, it is mainly a three-dimensional biophysical filter composed of proteoglycans (PGs) and glycosaminoglycans (GAGs), along with different kinds of fibers (e.g., collagen and elastin) and the gel-like ground substance. It is called the *matrix* and filters all substances moving from vessels to the cell and from the cell to vessels.

The ground substance derived its name from being a semifluid gel. Tissue fluid is loosely associated with the ground substance, thereby forming the medium for passage of molecules throughout supporting cells and for the exchange of metabolites with the circulatory system[15]. The terms extracellular matrix and ground substance are used in modern histology[16].

The matrix is synthesized by fibroblasts that react to damage of the matrix within minutes to restore its condition. Chondrocytes generate cartilage matrix, and osteoblasts generate bone matrix. Although the ECM is randomly secreted and does not conform to any mathematical principle, the same basic structure of PGs is seen overall in the ECM. Mainly, we see hyaluronic acid molecules carrying proteins in a treelike structure. Within this basic structure, we observe massively connected and free GAGs. The size of the PG structure is about 300 nm, leaving openings in the network structure of about 20–30 nm, through which interstitial liquid flows[17].

[15] Young B et al. *Wheater's Functional Histology*. 4th ed. Churchill Livingstone, Elsevier Science; 2000

[16] Young B et al. *Wheater's Functional Histology*, 4th ed. Churchill Livingstone, Elsevier Science; 2000

[17] Heine H. *Homotoxicology and Ground Regulation System (GRS)*. Baden-Baden: Aurelia Verlag; 2000

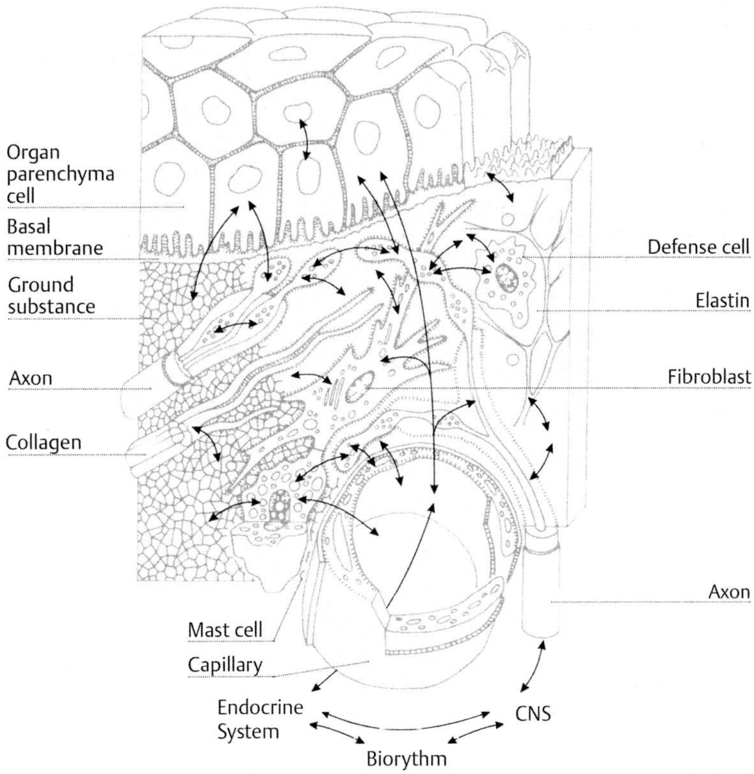


**Fig. 1.10** The ECM and Pischinger's ground substance in relation to cells and the extracellular space.

The electric potential of the ECM in the healthy condition is about 240 μV (microvolts). This potential increases in inflammatory conditions and decreases in degenerative conditions[18]. The matrix is negatively charged, attracting and holding positive ions and water. Thus, the matrix is hydrophilic, able to hold more than 1000 times its own weight in water. The gel-like ground substance, the main component of the matrix, is responsible for about 20% of our body weight (e.g., an adult weighting 80 kg has 16 kg of matrix).

The real terrain or "battlefield" of inflammation is the ECM. Although cells can be damaged by ECM inflammatory processes, the main action of inflammation is purely matrix related. There are various kinds of defense cells in the ECM; these cells observe the passage or presence of antigens and other toxins. In addition to macrophages, mast-cells are present to immediately degranulate upon immunoglobulin triggering, releasing histamine and phospholipids. In addition to T-helper cells, cytotoxic cells and natural killer (NK) cells also act in the ECM. Finally, neutrophils and eosinophils exist in the ECM and have phagocytic potential.

Under acidic conditions (e.g., during the first stage of inflammation), PGs and GAGs are broken down into small fragments by proteolytic and hydrolytic enzymes. These enzymes are secreted by fibroblasts, macrophages, and neutrophils. The

[18] Perger F. *Kompendium der Regulationspathologie und Therapie, der Formenkreis individueller Abwehrkraft.* Munich: Sonntag Verlag; 1990

residue of small fragments is absorbed by macrophages (by phagocytosis) and digested by lysosomal enzymes[19]. Deficiencies in the proteolytic and hydrolytic enzymes mentioned will result in corresponding storage diseases. Toxic substances and waste products may become trapped in the matrix structure, and deficient breakdown of the structure may cause extracellular depositions and intoxication. Thus, storage of toxic waste products so near to the cell may cause cell dysfunction in the short or long term, depending on the toxic and/or obstructive load of the substance.

In dealing with extracellular pathogens, the immune system aims to destroy the pathogen itself and to neutralize its products[20]. The immune system tries to eliminate pathogens in the extracellular space, before they get to the cell. Once the pathogen is in the cell, only cytotoxic activity remains to destroy the infected cell. Because many pathogens have an extracellular and an intracellular phase, extracellular elimination avoids intracellular intoxication. Again, the efficiency of the extracellular defense in the ECM is crucial for human health.

The matrix structure is continuously renewed. Enzymes induced by proinflammatory cytokines the so-called metalloproteinases, will break it down, whereas fibroblasts will build it up. Accordingly, efficient elimination of toxic substances that are trapped in the structure is possible. The benefit of a controlled inflammatory process is the elimination of toxins from the matrix to avoid extracellular accumulation and disturbance of the transport function of the biophysical filter itself.

The bidirectional arrows in **Fig. 1.10** illustrate the reciprocal interactions between different structures and ARSs, mediator flows, transport of substances, and reciprocal stimulation or inhibition.

### *Confirmation in Modern Histology*

The ECM with its ground substance is accepted as histological fact in modern medicine and is described in current medical textbooks[21,22]. The main difference between conventional Western medicine and complementary medicine is the understanding of the importance of the ECM when it comes to the development of diseases. Except for a few collagen-related diseases, the ECM has no major importance in disease evolution in conventional medicine and is certainly not the objective or even the terrain of action of its medications.

Within complementary medicine, the ECM plays a major role in the approach to disease evolution and in health. Because the viability of the cell is directly related to the purity of the ECM, the matrix is kept as uncontaminated as possible. Numerous therapeutic techniques and medications (drainage and detoxification) are used to cleanse and purify the ECM and maintain homeostasis.

### *Different Terminology, Same Topic*

ECM is now the standard term used to cover the direct environment of the cell. The main part of what Bernard called the internal environment and what Pischinger called the GRS is known today as the ECM. Pischinger's term ground substance still remains in use. The terminology used by Lamers et al. (BBRS, for basic bioregulating system) is not accepted in the worldwide scientific literature, though it is used in neural therapeutic literature. "Mesenchyma" (older terminology) and "connective tissue" are not synonyms for what is meant by the matrix in bioregulatory medicine. Connective tissue refers mainly to collagen and elastin, which are only a part of the ECM.

The structure of the connective-tissue fibers may play a significant role in the origin of disease. According to some authors, this tissue might be a

[19] Heine H. *Homotoxicology and Ground Regulation System (GRS).* Baden-Baden: Aurelia Verlag; 2000

[20] Roitt I et al. *Immunology.* 7th ed. Mosby Elsevier; 2006

[21] Young B et al. *Wheater's Functional Histology.* 4th ed. Churchill Livingstone, Elsevier Science; 2000

[22] Junqueira LC et al. *Basic Histology.* 11th ed. McGraw-Hill/Appleton & Lange; 2005

very subtle information and steering network in addition to the nervous system. Network cells communicate and interact with each other via this mechanosensitive communication. Although connective tissue is a new item in medical research, we may speculate that it has a much more important task than simply keeping the bodily tissues together.

### Detoxification of the Extracellular Matrix

Many foreign substances, benign or malign, will end up via the bloodstream in the ECM where they may be stored. In addition to the activity of defense cells at that level, most foreign toxins and metabolic waste products will be eliminated by the venous and lymphatic systems. This is the objective, which is not always reached. The same system that protects the cells from toxic burden may be the cause of long-term health problems. Everything has to do with the way the biophysical filter (i.e., the matrix) is constructed.

#### *The Proteoglycans and the Glycosaminoglycans*

The fine-meshed web of PGs (see **Fig. 1.11**) and GAGs creates a filtering system between the cell and the blood circulation and lymph system. This avoids direct presentation of toxic burden to the cell. On the other hand, the structure of the web might trap toxins, metabolic waste products, and other unwanted elements and might cause an obstruction of the flows of interstitial fluid toward and away from the cell. Thus, the supply of essential substances and the transportation of cellular waste products could be obstructed, impairing cell function.

#### *Trapping of Toxins*

**Figure 1.12** displays the three main routes by which toxins may become trapped in the ECM and cause deregulation of homeostasis: a mechanical route (a), an electric route (b), and a hydrophilic route (c). The fine web structure of the ECM, consisting mainly of PGs and GAGs, is responsible for this trapping.

**Mechanical Route**

A PG (as a single structure) is about 300 µm, measured at the level of its backbone core protein. Because of its transversal branchlike proteins (**Fig. 1.11**), the diameter of the openings in the mesh-like structure will be less than 30 µm. Consequently, any structure floating in the interstitial fluid might become trapped in the matrix simply because of its size.

Proteins are often bigger than 30 µm, so they are the substances that will most often be "stored" in the matrix and that may physically obstruct homeostasis. Because of the lifespan of this PG

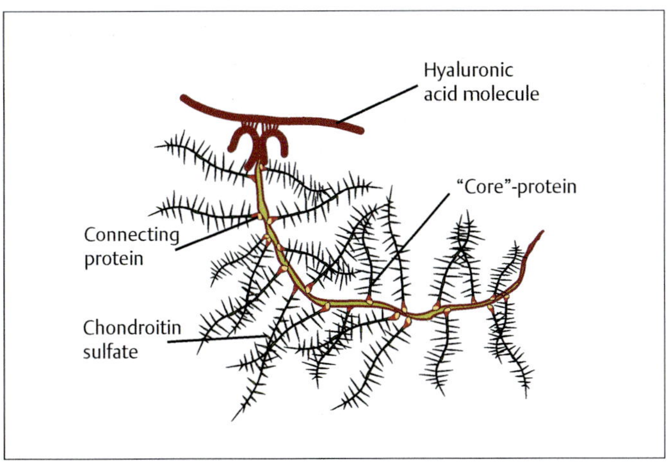


**Fig. 1.11** The fine-meshed web of PGs and GAGs.

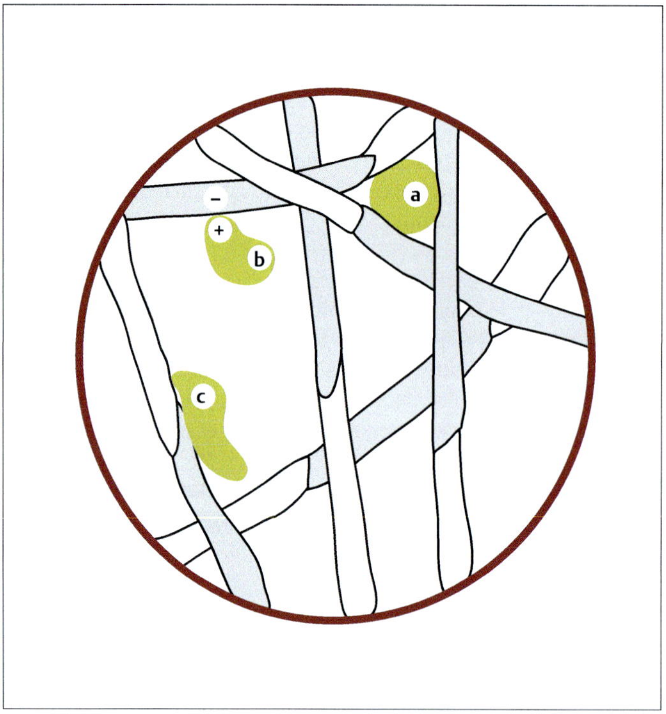


**Fig. 1.12** The three main routes by which toxins may cause deregulation of homeostasis. (**a**) a mechanical route, (**b**) an electrical route, (**c**) a hydrophilic route.

structure in the ECM, which is a maximum of 120 days before it is replaced by the fibroblasts present, toxins may "move" through the structure each time it is replaced. The first (acidic) phase of an inflammatory process will "dissolve" the PG structure, opening the gate for phagocytosing cells to eliminate the toxins when possible.

### Electric Route

The GAGs are acidic and therefore the electric potential of the PG/GAG structure in the matrix is negative. This means that positive ions will be attracted by the structure and are open to linking themselves to the structure because of its opposite charge. The opposite electrical charge of the GAGs and PG structure attracts positively charged toxins and causes them to remain in the structure.

### Hydrophilic Route

The PG/GAG structure is also very hydrophilic because of the negative charge, and it attracts mainly sodium ions. This means that the structure "holds" water: of the mass of the ECM, 10% is composed of GAGs and 90% is water[23]. Water-soluble toxins may become trapped through these hydrophilic properties of the PG/GAG structure because it is so rich in water.

### *Inflammation as a Turbocleansing Process*

Fibroblasts in the extracellular space continuously secrete the PG/GAG matrix, keeping the biophysical filtering function optimized. If tissue is damaged (e.g., a wound) or if the load of toxins, metabolic waste products, or any form of antigens in the ECM is increased, an effective defense system will trigger inflammation, producing an acidic stage of the extracellular environment.

During inflammation, all kinds of proteolytic and hydrolytic enzymes break down the fine-meshed web of PGs and GAGs, leaving a gel-like liquid behind. Toxins trapped in the matrix are set

[23] Alberts B et al. The extracellular matrix of animals. *Molecular Biology of the Cell.* 4th ed. New York: Garland Science, New York; 2002:1090–1113

free and can undergo phagocytosis or be transported via the lymphatic system to the bloodstream, where they are finally filtered out by the liver.

In alkaline stages of the process, fibroblasts rebuild the structure, providing the extracellular space with its biophysical filter again. Fibroblasts can generate ground substance within minutes[24].

Inflammation is a "turbocleansing" process, temporarily changing the structure of the ECM so that toxins, metabolic waste products, and antigens within it can be eliminated.

## Inflammation as an Autoregulating System

Inflammation is a complex process of defense in which many different immune cells are mobilized and involved. Many mediators, including those released by endocrine cells, tissue cells, and defense cells, are used to steer the actions of these cells. For each agonist pushing a subpathway forward, there are one or more antagonists seeking to exert the opposite effect. Inflammation is a complex masterpiece of autoregulation.

This complexity becomes even greater once we consider the interactions between the inflammation cascade itself as an autoregulating process and other autoregulating processes in the body, such as the hypothalamus–hypophysis–suprarenal axis, thermoregulation, and vasomotor regulation. All of these processes interact with the inflammation-regulating defense system. External factors may also strongly influence the inflammatory outcome, making the autoregulating process even more complex.

### Biological Objectives of Inflammation

From a bioregulatory viewpoint, inflammation is not seen just as a sign of disease, showing symptoms such as pain, swelling, warmth, redness, and loss of function of the affected tissue. Behind the mobilization of defense cells, we recognize the biological objectives of the process: the elimination of antigens and toxins and the restoration of bodily tissues. Homeostasis is the goal. Although the patient's quality of life is the first priority, we should see this in a long-term perspective. Within homotoxicology, the short-term suppression of a matrix cleansing process, such as inflammation, is acceptable only in life-threatening circumstances. Bioregulatory medicine is used to intervene in the inflammatory process in such a way that its biological objectives remain unhindered but relief of the symptoms is realized in the short term. Blocking off or suppressing enzymatic processes, as is done with traditional nonsteroidal anti-inflammatory drugs (NSAIDs) or corticoids, is certainly not an option in bioregulatory medicine.

Once inflammation is understood as a cleansing process, mainly of the ECM (being cell and tissue protective), it will be dealt with differently. Suppression of symptoms will no longer be the first priority. We will no longer block enzymatic processes that lead to the suppression of indispensable steering mediators as a standard therapy. We seek therapeutic measures that are in full compliance with the inflammation process by using medications that will regulate the process within the margins acceptable to the patient (symptomatic relief) without disturbing the biological objectives of the process. This means that we use bioregulation therapy to bring the wave pattern and oscillations of the inflammation autoregulating process within a more purposeful range around the set point.

### Mediators in Microdosages and Nanodosages

Cytokines have well-defined tasks within the inflammatory process. In the action of cytokines, receptors are able to translate a docking cytokine into a specific cell action[25]. It is notable that the organism uses very subtle dosages of mediators to steer the inflammation process, sometimes causing "exponential" effects. Concentrations of cytokines, such as interferons (IFNs), interleukins (ILs),

[24] Lozzo RV. Biology of Disease. Proteoglycans: Structure, Function, and Role in Neoplasia. *Lab Invest.* 1985;53:337–396

[25] Thomson AW et al. eds. *The Cytokine Handbook.* 4th ed. London: Academic Press; 2003

chemokines, and other inflammation-steering mediating factors in the body, are within the range of micrograms and nanograms per milliliter. In the blood plasma, the range is between nanograms and picograms per milliliter.

Dosages are so minute that the enzyme-linked immunosorbent assay (ELISA) method is often needed to mark the values. The following values were detected in a study of a population of 30 healthy plasma donors: IL-2, 0–74.48 pg / mL; IL-4, 0–65.18 pg / mL; IL-5, 0–86.69 pg / mL; IL-6, 0–73.17 pg / mL; IL-8, 0–74.8 pg / mL; IL-10, 0–12.6 pg / mL; IL-13, 0–3.9 pg / mL; and transforming growth factor (TGF) $\beta_1$, 14.98 ± 12.39 ng / mL.

These low dosages of mediators, which can often generate huge responses of the system, are characteristic in autoregulating processes in living systems. They contrast with the macrodosages used in conventional medications to intervene in the same systems.

A subtle, fragile system like a bodily ARS, able to deviate from its set point at the smallest pulse of a mediator or stimulus, should be approached with subtle dosages of medications, intervening in the circuit to avoid inducing rigidity of the system itself. Rigid ARSs are seen within bioregulatory medicine as one of the main causes of chronic diseases.

### *Effects of Conventional Anti-inflammatory Medications*

Nonsteroidal anti-inflammatory drugs block the cyclooxygenase (COX) pathway, decreasing the synthesis of proinflammatory prostaglandins. There are nine known prostaglandin receptors on various cell types. Prostaglandins can thus act on a variety of cells through these receptors. The varied actions of prostaglandins include the following:

- They cause constriction or dilation in vascular smooth-muscle cells (i.e., vasoconstriction or vasodilation, respectively).
- They create constriction of smooth muscles.
- They intervene in blood platelet aggregation.
- They sensitize spinal neurons to pain.
- They are known for their regulating effect in inflammatory mediation.
- They regulate calcium movement and intervene in hormone regulation.
- Finally, they are involved in controlling cell growth.

Other bodily mediators also have an isolated effect in inflammation or a very varied simultaneous effect on many immune and tissue cells at the same time. Tumor necrosis factor (TNF) $\alpha$is another example; its direct effects include the following.[26]:

- It has a prothrombotic effect because it promotes leukocyte adhesion and migration.
- It plays an important role in the regulation of macrophage activation.
- It has an important role in triggering immune responses in tissues.
- It modulates hematopoiesis and lymphocyte development.
- It triggers nuclear transcription factor (NF-κB) via receptors on the cell membranes, resulting in selective gene activation and consecutive cell action and secretion.

Many other mediators have multiple simultaneous effects, changing parameters within the inflammation cascade. Interactions are very complex; changes of activity of one or more mediators, induced by macrodosages of conventional medications, will rarely have an isolated effect because of the varied effects created by the influenced mediators themselves.

Regarding the effect of NSAIDs, heavy inhibition of prostaglandin synthesis has an octopus effect on many regulating systems in the body, mainly disturbing homeostasis and thus making the biological objectives of the inflammatory process almost unattainable. Although rapid symptomatic relief is often obtained, causal antigens, tissue damage, or toxins will remain longer as proinflammatory mediators and immune cell activities are strongly inhibited or even blocked.

When pathways of an ARS are blocked, other pathways try to compensate for the loss. In an inflammatory process, this means that the loss of the COX pathway through suppression by NSAIDs will be compensated by the effects of increased lipooxygenase and/or the thromboxane pathway. In fact, proinflammatory mechanisms compensate

---

[26] Roitt I et al. *Immunology*. 7th ed. Mosby Elsevier; 2006

for the loss of other proinflammatory mechanisms. These compensating actions cause the side-effects seen in the use of many NSAIDs and are also responsible for many contraindications mentioned with these medications. Increased lipooxygenase as a consequence of selective COX inhibition may trigger bronchial spasms, a reason for contraindication against use of NSAIDs with bronchial asthma.

Similar effects on the thromboxane pathway may occur because of use of COX-2. The COX-2 inhibitors may have cardiovascular risks. Because COX-2 inhibitors cause a decline in prostacyclin without any change in thromboxane levels, a relative dominance of thromboxanes is created in the inflammatory balance; with this there are higher risks of heart attacks and strokes, especially in patients with higher blood pressure. Although COX-2 inhibitors seem to have a gastrointestinal benefit compared with conventional NSAIDs[27], many epidemiological studies have supported and amplified concerns about the risk of myocardial infarction and serious cardiovascular events associated with rofecoxib[28], and other COX-2 inhibitors.

The COX-2 inhibitors deregulate agonist/antagonist (prostaglandin-$E_2$/thromboxane) effects in the inflammatory autoregulating process, with a higher risk of side-effects as a result. It is not really a solution to use acetylsalicylic acid (ASA) along with a COX-2 inhibitor to dilute the blood and counter thromboxane-induced side effects. Acetylsalicylic acid is, in fact, a conventional NSAID (a COX-1 and COX-2 inhibitor) and, thus, causes gastrointestinal disorders. (The goal of using COX-2 inhibitors is to avoid gastrointestinal disorders.)

Autoregulating systems will always seek or incline toward measures that allow them to realize their biological objectives. The higher the suppressing stimulus on one or more subpathways from the outside, the stronger the compensating reaction from other unaffected subpathways belonging to the same ARS to reach the general objective (homeostasis). Blocking subpathways with conventional medications will rarely be without side-effects, resulting from the increased activity of compensatory subpathways. In fact, the use of suppressive medication is the fastest way to overcompensate even for rigid ARSs.

## *The Arndt–Schulz Rule*

The Arndt–Schulz rule applied to pharmacological effects is named after Rudolf Arndt and Hugo Paul Friedrich Schulz. It is better to speak about a rule and not about a law, as is seen in some literature, because although the rule fits for many substances there are also many exceptions.

> ***The Arndt–Schulz rule***
>
> Small doses of a substance stimulate, moderate doses of the same substance inhibit, and large doses stop physiological processes (or even kill).

High acute dosages of cortisone will kill cells, whole tissues, or even the patient (lethal doses). In moderate doses, cortisone has a strong multiple anti-inflammatory effect, and conventional medicine uses moderate dosages of cortisone in highly inflammatory reactions. However, the moderate physiological effect of cortisone is inverted at small or minute dosages. Research has shown its highest inverse effect in the concentration range of $10^{-9}$, or 1/1 000 000 000.

High acute concentrations of alcohol may be lethal. Moderate concentrations cause a drunken state and have a sedative effect. Low concentrations create an awakening or stimulating effect.

This inverse dependence of physiological effects on minute doses makes it possible to develop medications with minute concentrations of an active component. *Arnica montana*, a plant, member of the Asteraceae family, is toxic and in very high concentrations may cause death. At moderate doses the helenalin in *A montana* is responsible for irritation and inflammatory symptoms in the mucous membranes. It may affect in pulse rate,

[27] Bombardier C et al. Comparison of upper gastrointestinal toxicity of rofecoxib and naproxen in patients with rheumatoid arthritis. *N Engl J Med.* 2000;343:1520–1528

[28] Topol EJ et al. A coxib a day won't keep the doctor away. *The Lancet.* 2004:364:639–640

cause palpitations of the heart, and cause muscle paralysis[29].

Small dosages of *A montana* have curative effects without being toxic. Research has indicated that helenalin and dihydrohelenalin, two of the main active esters in *Arnica*, interfere with various metabolic inflammatory processes. These parameters are effectively influenced at $10^{-4}$ to $10^{-6}$ mol/L concentrations. These effects were similar to those found with the use of conventional NSAIDs, which are used at much higher molar concentrations. *Arnica* has proved to have anti-inflammatory, antirheumatic, and antiarthritic effects[30,31]. In-vitro research has shown that both esters also have an inhibiting effect on the histamine release from mast cells. Basic research has also shown serotonin secretion from human blood platelets with simultaneous inhibition of thromboxane $B_2$ formation. Both components of *Arnica* may indirectly interfere with phospholipase $A_2$ activity.[32]

Other phytocomponents showed similar inflammation-regulating effects in minute dosages[33]. Three isorhamnetin glycosides from *Calendula* flower showed inhibition effects in vitro (at concentrations of $1.5 \times 10^{-5}$ mol/L) on arachidonate 12-lipoxygenase from rat lung cytosol[34]. In-vivo immunomodulatory effects were observed with minute dosages of *Calendula officinalis*. Polysaccharide fractions from aqueous extracts of *Calendula* flower appeared to enhance phagocytosis in the carbon clearance test in mice[35]. The unsaponifiable fraction of a hydroalcoholic extract, given intraperitoneally, protected mice from sepsis due to *Escherichia coli* [36].

### *Inflammation-regulating Drugs and Other Immunomodulators*

Immunomodulators are medications that intervene to regulate one or more parts of the immune system. The inflammation-regulating drugs (IRDs) are a new subclass within the family of immunomodulators and are discussed in detail in Chapter 4 in terms of immunomodulation as one of the three pillars of anti-homotoxic treatment. In contrast to conventional NSAIDs, the IRDs do not block enzymatic processes within the inflammatory cascade. The IRDs have a regulating effect, mainly induced at the top of the inflammatory cascade in antigen-induced inflammatory conditions and consequently influencing all of the subsequent mediators lower in the chain. In non–antigen-induced inflammation, tissue cell damage often occurs, releasing phospholipids. Inhibition of proinflammatory mediators such as IL-1β, TNF-α, and even IL-8, cases down-regulation of the inflammatory process[37] at sites other than the cyclooxygenase pathway (explained in detail later). Thus, IRDs often have multiple sites of action, influencing the whole inflammatory cascade.

Because IRDs down-regulate excessive expression of defense by inhibiting intracellular messengers, and not by blocking enzymatic processes or occupying essential receptors, the autoregulatory steering mechanism remains unchanged with use of IRDs. Because all activities lower in the inflammation cascade than the sites of action of the drug are down-regulated to the same extent, compensatory activities of the immune system (seeking

[29] Barnes J et al. *Herbal Medicines*, 2nd ed. London: Pharmaceutical Press; 2002

[30] Willuhn G et al. *Phytomedicines of Europe. Chemistry and Biological Activity.* ACS Symposium series 691. Washington DC: American Chemical Society; 1998: 118–132

[31] Von Raison J et al. Arnika—Arzneipflanze mit Tradition und Zukunft. *Z Phytother.* 2000;21:39–54

[32] Schröder H et al. Helenalin and 11 alpha,13-dihydrohelenalin, two constituents from *Arnica montana* L., inhibit human platelet function via thiol-dependent pathways.*Thromb Res.* 1990;57:839–845

[33] ESCOP Monographs. *The Scientific Foundation for Herbal Medicinal Products*, 2nd ed. Stuttgart: Thieme; 2003

[34] Bezakova L et al. Inhibitory activity of isorhamnetin glycosides from Callendula officinalis on the activity of lipoxygenase. *Pharmazie.* 1996;51:126–127

[35] Wagner H et al. Immunstimulierend wirkende Polysaccharide (Hetroglykane) aus höheren Planzen. Arzneimittelforschung. 1985;35:1069–1075

[36] Delaveau P et al. Drogues végétales stimulant l'activité phagocytaire du système réticulo-endothélial. *Planta Med.* 1980;40:49–54

[37] Porozov S et al. Inhibition of IL-1β and TNF alpha secretion from resting and activated human immunocytes by homeopathic medication Traumeel S. *Clin Dev Immunol.* 2004; 11(2):143–149

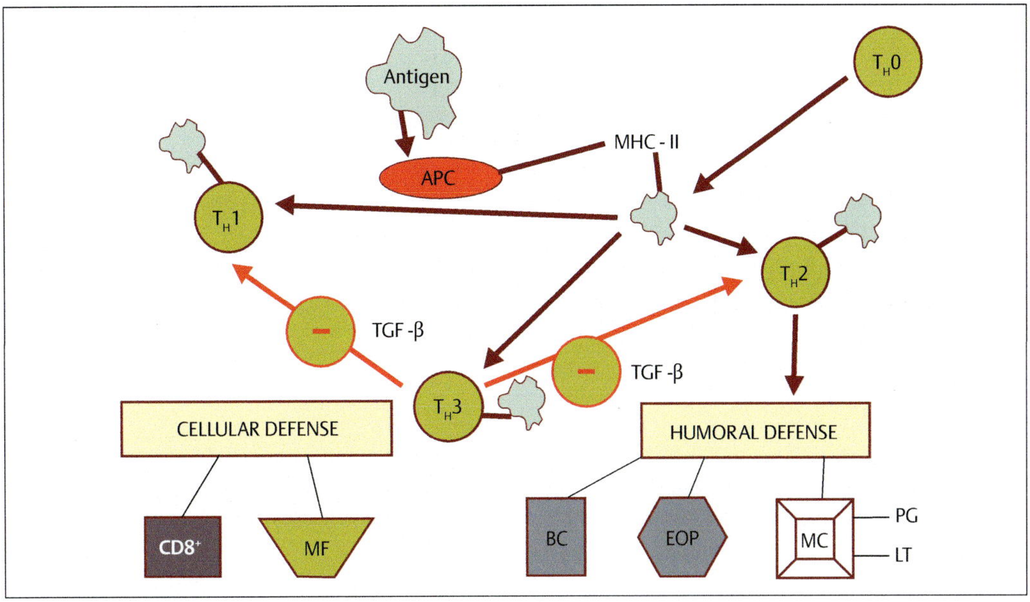


**Fig. 1.13** The down-regulating effect of TGF-β.

compensation of lost or blocked pathways) are not seen. This is the main reason why IRDs have essentially no reported side-effects and are contraindicated in very few pathological conditions.

Of all the IRDs available worldwide, Traumeel from the pharmaceutical firm Heel is probably the best known and most prescribed. Traumeel has been shown to stimulate the production of specific T-regulatory ($T_{reg}$) cells ($T_H3$ $CD4^+$ cells) that by the release of TGF-β inhibit proinflammatory $T_H1$ and $T_H2$ lymphocytes[38,39]. TGF-β is an anti-inflammatory mediator and has a fibroblast-enhancing effect, resulting in faster ECM damage repair and thus also tissue repair. Traumeel therefore has a wound healing effect in addition to its inflammation-regulating capacities.

**Figure 1.13** shows this down-regulating effect of TGF-β at the top of the inflammation cascade in antigen-induced inflammation. Minute concentrations of components of Traumeel will be taken up by antigen-presenting cells (APCs), such as dendritic cells, and will present characteristic proteins via the major histocompatibility complex (MHC) class II or HLA class II of the APC. Once prolymphocytes ($T_H0$ cells) have responded to this motif information, they become $T_{reg}$ cells, secreting TGF-β once they come in contact in the inflammation terrain with $T_H1$ and $T_H2$ cells with similar antigen-induced motif information. The release of TGF-β will inhibit the activity of the proinflammatory lymphocytes, resulting in a down-regulated inflammatory process. (also see Chapter 4 for a more detailed approach). Other recent research[40] shows a clear down-regulation of proinflammatory cytokines (e.g., IL-1β, TNF-α, and IL-8) by Traumeel.

Other immunomodulators, such as Engystol, show a similar phenomenon, intervening at the level of the release of steering mediators. Recent research[41] shows an increase of 24% of IFN-γ

[38] Heine H et al. Immunologische Beistandsreaktion durch pflanzliche Extrakte in antihomotoxischen Präparaten. Biologische Medizin. 1998;27(1):12–14

[39] Heine H. et al. The anti-inflammatory action mechanism of an anti-homotoxic composite remedy. *Ärtzezeitschrift für Naturheilverfahren.* 2002;43:2

[40] Porozov S et al. Inhibition of IL-1β and TNF alpha secretion from resting and activated human immunocytes by homeopathic medication Traumeel S. *Clin Dev Immunol.* 2004;11(2):143–149

[41] Enbergs H. Effects of the homeopathic preparation Engystol® on interferon-γ production by human T-lymphocytes. *Immunol Invest.* 2006;35:19–27

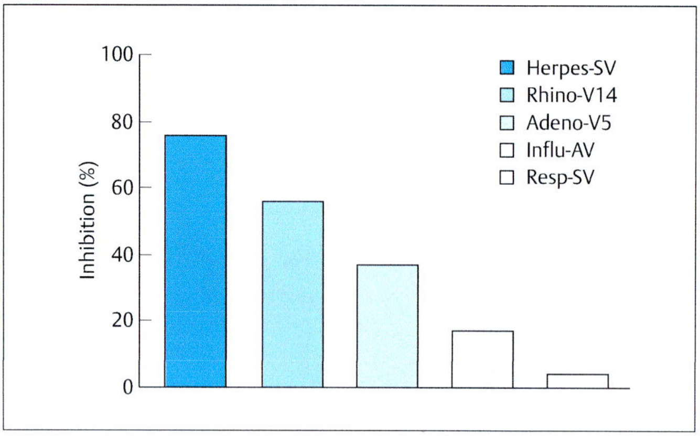


**Fig. 1.14** Inhibition in percent of the proliferation of different viruses by Engystol. In vitro study, 2005.

secretion by $T_H1$ lymphocytes because of Engystol. Other research has shown that Engystol reduces the proliferation of various viruses, as **Fig. 1.14** illustrates, probably because of its effect of increasing IFN-γ secretion[42].

Both examples of immunomodulating medications used in bioregulatory medicine, Traumeel and Engystol contain only minute concentrations of active components, resulting in strong immunomodulating effects, as studies have shown[42]. Minute concentrations of active components have bee shown to have physiological effects, in strong contrast to what conventional pharmaceutical thinking might suggest or even proclaim. Minute dosages of the right protein-containing substance, presented to the immune system, may have a strong immunomodulating effect.

## The Greater Defense System

The greater defense system[43] consists of six subsystems working together to eliminate toxins from the body and to restore tissue damage. Detoxification and drainage as well as immunomodulation are the main pillars in bioregulatory treatment and certainly in antihomotoxic treatment.

*The Six Subsystems of the Greater Defense System*

1. The reticuloendothelial system
2. The hypothalamus–hypophysis–suprarenal axis
3. The neural reflex system
4. Detoxification by the liver
5. Detoxification of the matrix
6. The mucous membranes.

These systems all merge into each other and have a synergetic effect, working together to defend the body and maintain homeostasis.

### *The Reticuloendothelial System (RES), or the Mononuclear Phagocyte System*

The RES is a part of the immune system, consisting of phagocytosing cells, mainly macrophages, located in reticular connective tissue; monocytes, located in the bloodstream; and dendritic cells, mainly located in the epithelial surfaces. These cells are stored and accumulated in lymph nodes and the spleen; they are ready to act if there is any invasion of antigens. Variations of these cells, such as histiocytes in tissues and Kupffer cells in the liver, are also part of the RES.

The terms *lymphoreticular system* and *mononuclear phagocyte system* are both synonyms for the RES. Some authors have proposed that the term mononuclear phagocyte system should be

---

42 Oberbaum M, Glatthaar-Saalmuller B, Stolt P, Weiser M. Antiviral activity of Engystol: an in vitro analysis. *J Altem Complement Med.* Oct 2005;11(5):855–862

43 Reckeweg H-H. *Homotoxicology, Illness and Healing through Anti-homotoxic Therapy.* 3rd edi. Menaco Publishing Company, Aurelia Verlag; 1991

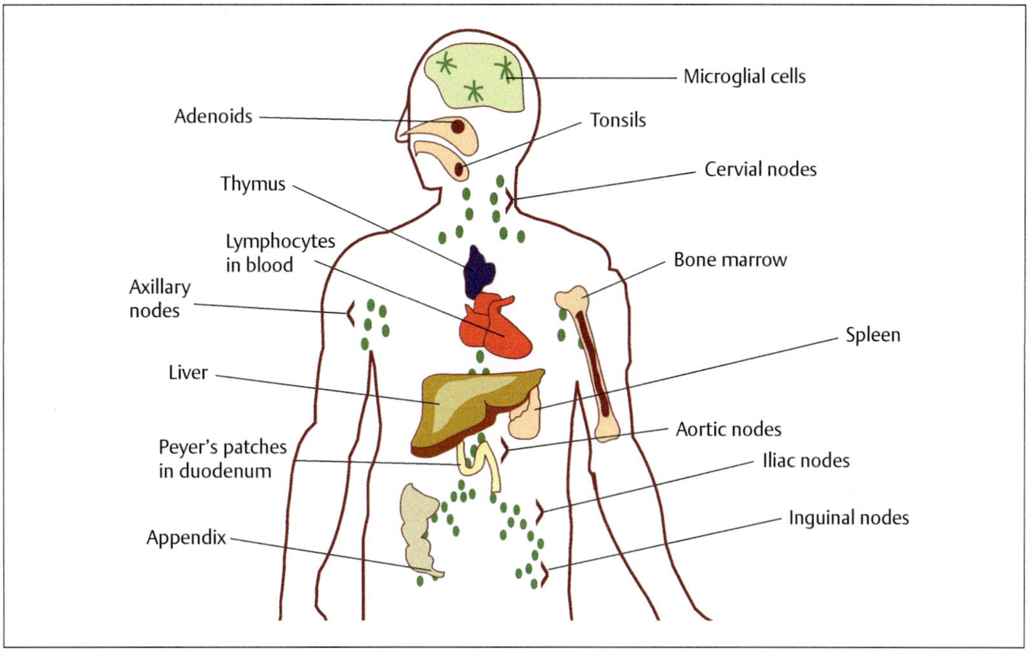


**Fig. 1.15** The tissues of the reticuloendothelial system.

adopted rather than the almost universally used term RES.

The RES is divided into primary and secondary lymphoid organs. The primary organs refer to the location of origin of RES cells: the bone marrow and the thymus. The secondary lymphoid organs refer to the location where the RES cells really act. This includes the encapsulated lymph tissue (lymph nodes), the spleen, and the mucosal-associated lymphatic tissue (MALT), mainly divided into a gastrointestinal part (gut-associated lymphatic tissue: GALT) and a respiratory part (BALT: bronchial-associated lymphoid tissue). **Figure 1.15** provides an overview of the RES tissues.

The activity of the RES is directly related to the presence of antigens and denatured tissue in the ECM and mucous membranes. Deficiency of the RES will result in accumulation or storage of toxins and burden in the ECM structure, endangering cell function first and survival of the cell later. Part of RES activity is also the formation of immunoglobulins, initiated by the activity of dendritic cells and other APCs.

### *Dendritic Cells and Motif Formation*

Dendritic cells were discovered by Steinman and his team in 1973[44]. Today, they are known to be a small population of immune cells occurring in the body surfaces, where the organism is in direct contact with the environment (the skin's Langerhans cells and the mucosal layers). They are continuously seeking antigens of which they will present characteristic proteins to other immune-competent cells; this is why they belong to the group of APCs.

Once activated by antigen contact, dendritic cells migrate to the nearest lymph nodes, where they transmit the characteristic antigen information to T- and B-lymphocytes. Through this, a specific immune response against the antigen is induced[45].

---

[44] Steinman RM et al. Identification of a novel cell type in peripheral lymphoid organs of mice. I. Morphology, quantitation, tissue distribution. *J Exp Med.* 1973;137(5):1142–1162

[45] Sallusto F et al. The instructive role of dendritic cells on T-cell responses. *Arthritis Res.* 2002;4(Suppl 3): S127–132

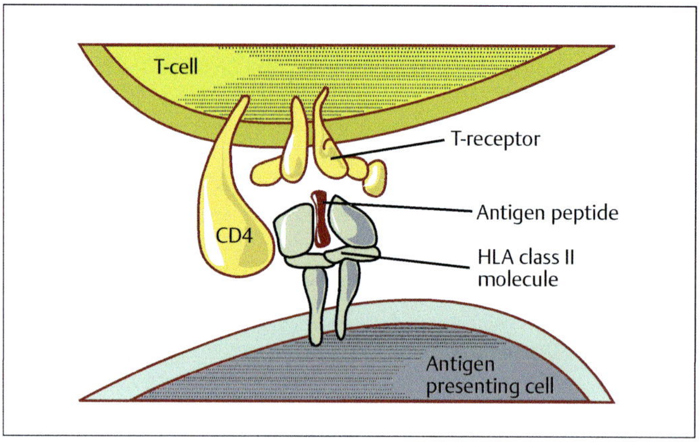


**Fig. 1.16** Presentation of a specific antigen peptide in the HLA class II (MHC class II) of the APC. This information is transmitted to the T-helper cells via the T-cell receptor.

As **Fig. 1.16** illustrates, a specific antigen peptide is presented in the HLA class II (MHC class II) of the APC. This information is transmitted to the T-helper cells via the T-cell receptor; the T-helper cell will induce cellular or humoral defense against the antigen.

Proinflammatory $CD4^+$ cells need antigen peptide activation (the antigen-specific motif presented by the MHC or HLA of the APC) to become specific T-lymphocytes steering the specific defense (immunoglobulin production). In this way, the specific pathway of the inflammation reaction of the organism is steered against the antigen. The nonspecific pathway of the defense system does not require motif recognition.

Certain immune-competent cells, such as natural killer cells, react directly to damaged cells or virally infected cells, whereas others are induced by another class of MHC, the MHC type I, which will induce T-cytotoxic cells of the $CD8^+$ type.

*Antigen-specific Antibody Formation*

The antigen specificity of an antibody, produced by B lymphocytes, is extremely precise. Antibodies are exceptional in their almost unlimited potential for diversity, and for each different antigen the organism will try to create a specific antibody that is not effective for any other antigen[46].

Most antigens will induce T- and B-cell recognition. B-cells are activated, proliferate, and differentiate via an MHC class II motif presentation by T-helper cells. This leads to the generation of plasma cells (immunoglobulin production) and memory B-cells. These antigens are called T-dependent antigens. A small group of antigens, however, is able to activate B-cells directly; this means the cells are activated without MHC class II–activated T-cells and are, therefore, referred to as T-independent antigens[47].

There are five main isotypes of antibodies, some with subclasses:

- **IgA:** This immunoglobulin is found in the same areas in which dendritic cells are mainly found, such as the mucosal layers of the gut and the respiratory and urogenital tracts. IgA prevents the colonization by pathogens in these regions. This antibody type is also found in saliva, tears, and breast milk.
- **IgD:** This immunoglobulin functions mainly as an antigen receptor on B-cells. Its function is less well defined than that of other isotypes.
- **IgE:** This immunoglobulin is a well-known immune component in allergies. It binds to allergens and triggers histamine release from mast-cells. IgE also protects against parasitic worms.
- **IgG:** In its four forms (IgG1–IgG4), this immunoglobulin provides the majority of antibody-based immunity against invading pathogens.

[46] Roitt I. et al. Immunology, 7th edition, Mosby Elsevier, 2006

[47] Roitt I. et al. Immunology, 7th edition, Mosby Elsevier, 2006

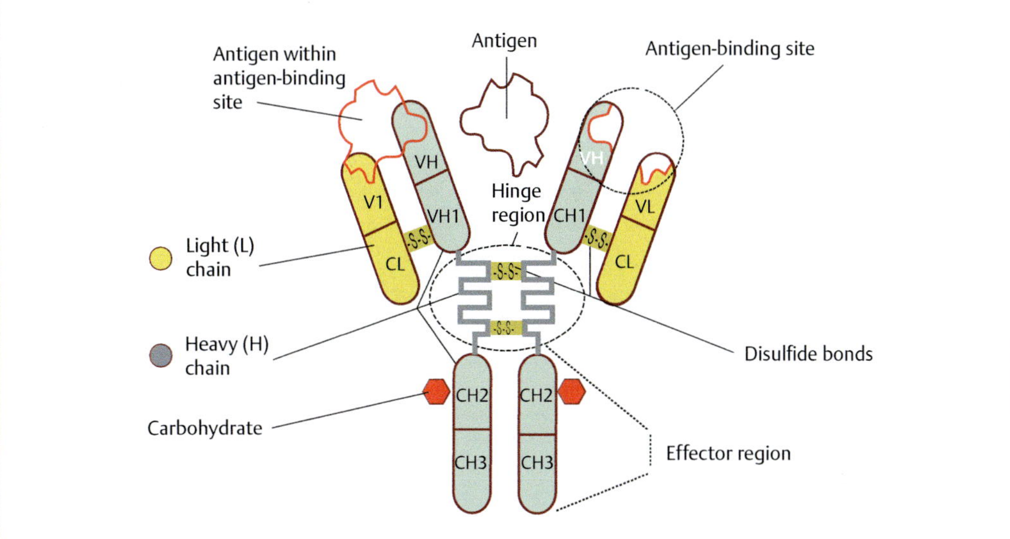


**Fig. 1.17** Schematic representation of the structure of an antibody: a standardized heavy chain with a variable antigen-specific light chain. With this arrangement the same basic material can produce antibodies against a hugely diverse set of antigens.

- **IgM:** This immunoglobulin eliminates pathogens in the early stages of $T_H2$ pathway-mediated (humoral) immunity, before there is sufficient IgG.

As shown in **Fig. 1.17**, each antibody has a standardized heavy chain on which an antigen-specific light chain is bound. This variable light chain on a standard heavy chain makes it possible to use the same basic material for antibody production for a diverse set of antigens[48].

### *The Hypothalamus–Hypophysis–Suprarenal Cortex System*

This system has a regulating hormonal steering effect, directing defense mechanisms during stress-induced processes such as inflammation. As mentioned earlier, it is a magnificent example of an ARS using negative feedback to regulate release of hormones. The corticotropin-releasing hormone (CRH) is negatively influenced by the release of cortisol by the suprarenal cortex, resulting in lower cortisol release at the end of the loop.

A good example of a bipolar feedback system in humans is given by the fact that the hypophysis can influence inflammation in an inhibitory or a stimulatory way. This existence of opposite steering capacities of the same gland was hypothesized by the Hungarian researcher Hans Selye and was confirmed in later research[49,50]. In 1949, Selye discovered that an inflammatory reaction that can be induced in the rat by the parenteral administration of egg white is inhibited by cortisone or purified ACTH. On the other hand, desoxycorticosterone acetate, a mineral corticoid compound, tends to aggravate the reaction.

[48] Pier GB et al. *Immunology, Infection, and Immunity.* Washington, DC: ASM Press; 2004

[49] Berczi I. Immunoregulation by pituitary hormones. In: Berczi I, ed. *Pituitary Function and Immunity.* Boca Raton, FL: CRC Press; 1986

[50] Berczi I. The role of the growth and lactogenic hormone family in immune function. *Neuroimmunomodulation.* 1991;1:201–216

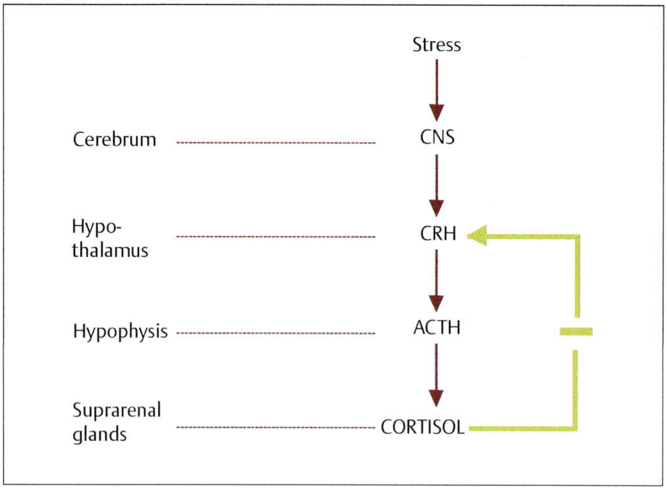


**Fig. 1.18** Effects of stress on the hypothalamus–hypophysis–suprarenal system. Note the negative feedback loop in which excess cortisol activates the brain's glucocorticoid receptors and suppresses the production of CRH.

*Psycho-neuro-endocrine Immunology*

Psycho-neuro-endocrine immunology (PNEI) is a new branch in medicine in which the interaction between psychological processes and the nervous, immune, and endocrine systems of the human body is studied. As described later, stress factors of any kind (perhaps especially emotions) might influence the efficacy of the immune system and ultimately affect tissues.

A short-term stress situation will induce the release of epinephrine from the suprarenal cortex. Prolonged stress factors, however, will induce the long-term release of cortisol, another "fight" hormone, with disastrous effects on tissues and the immune system after continuous exposure to this stress[51].

The spiritual father of "stress," Selye showed the influence of stress on thymus involution[52]. Modern research has shown remarkable results of tissue damage and immune deficiency with continued stress. Immune deficiency after continued increased levels of cortisol is confirmed in other research[53].

In normal conditions and healthy persons, when someone experiences a stressful event the level of glucocorticoids in the blood increases. The hypothalamus is activated via specific receptors in the hippocampus, and then secretes CRH. The CRH, in turn, causes the pituitary gland to release adrenocorticotropic hormone (ACTH) into the bloodstream. The ACTH then enters the adrenal glands and causes them to secrete cortisol. This process creates a negative feedback loop in which the excess cortisol activates the brain's glucocorticoid receptors and suppresses the production of CRH, as shown in **Fig. 1.18**.

Some authors even relate prolonged increased cortisol levels to the generation of a depressive state. The loop in **Fig. 1.18** (cortisol–CRH) no longer works, which will result in excess production of CRH and, hence, of cortisol. Many patients with serious depression have high blood levels of cortisol, caused by chronic stress. Modern research shows a possible negative influence of cortisol on available serotonin levels.

---

51 Henry JP. Biological basis of the stress response. *News Physiol Sci.* 1993;8:69–73

52 Selye H. Thymus and adrenals in the response of the organism to injuries and Intoxication. *Br J Exp Pathol.* 1936;17:234

53 van der Pompe G et al. An exploratory study into the effect of exhausting bicycle exercise on endocrine and immune responses in post-menopausal women: relationships between vigour and plasma cortisol concentrations and lymphocyte proliferation following exercise. *Int J Sports Med.* 2001;22(6):447–453

*The Diurnal Rhythm of Desoxycortisol and Cortisol*

Autoregulating processes often respond to a circadian rhythm, showing inverse levels of their intervening mediators in the day–night rhythm. Chronobiology studies this new field of interest within biology. In mammals, the "master clock" resides in the suprachiasmatic nucleus of the hypothalamus.

The day–night rhythm of the cortisol level is termed its *diurnal rhythm*. As indicated, cortisol release is directly related to ACTH release. **Figure 1.19** illustrates the peaks and troughs of ACTH and cortisol levels over 24 hours.

External triggers will often influence the release of a hormone. Dimming daylight will induce the secretion of melatonin, preparing the organism for sleep For example, research on melatonin and cortisol levels showed an inverse correlation in day–night rhythm. Cortisol has its peak of concentration in the morning, preparing the organism to be awake and alert for the day and for fight if necessary. The lowest 24-hour values will normally be registered between midnight (00:00) and 04:00, exactly when the deepest sleep occurs. Emotional stress and stress related to physical danger will influence the level of cortisol. If the level of cortisol remains high because of continued stress, the diurnal rhythm is disturbed.

Desoxycortisol can be considered as the mirror image of cortisol when it comes to its activity and levels: desoxycortisol increases when cortisol decreases and vice versa. Desoxycortisol has its main activity at night, stimulating the organism to detoxify, and if necessary even triggering inflammation to do so.

*Stress and Stress-related Exhaustion of ARSs*

In medical terms, *stress* can be defined as a deregulation of homeostasis, which is caused by physical or psychological stimuli.

**Hans Selye and the General Adaptation Syndrome**

Selye was the first researcher to demonstrate the negative impact of extended stress on the organism. Although most of his experiments were done on animal models, his extrapolations to human beings are accepted in modern science. He stressed the importance of the hypothalamus-hypophysis-suprarenal glands axis on immune dysregulation and its consequences in creating all kinds of diseases.

Selye posited the general adaptation syndrome (GAS) as a uniform model of dealing with stressors in mammals. This process of dealing with the

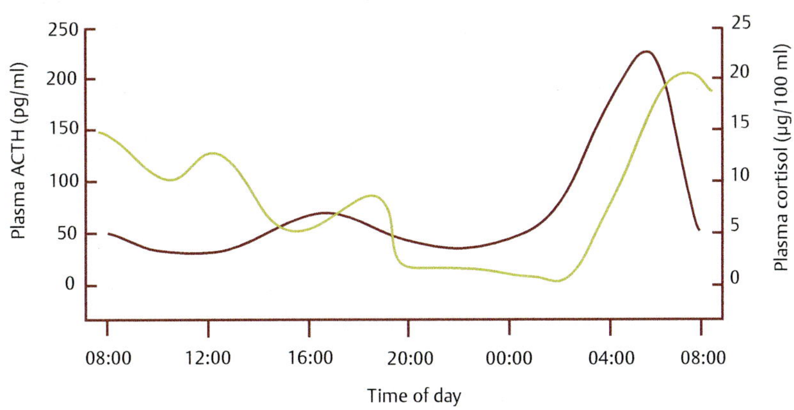


**Fig. 1.19** The diurnal rhythm of ACTH (heavy line) and cortisol (light line) levels over one 24-hour cycle.

stressors has three stages: the *alarm stage*, the *reaction stage*, and the *exhaustion stage*[54].

### *The Alarm Stage*

If a new stimulus appears, the organism immediately exhibits an alert state. The organism compares the stressor's demand with the opportunities of the organism itself. This means that stimuli that do not "threaten" will cause no stress at all. The organism deals with them "as usual."

However, if the stimulus threatens the organism, likely forcing it to change or adapt, epinephrine is released. Epinephrine will make the organism immediately ready for fight, flight, or freeze reactions. Epinephrine is released in the first stage, followed by cortisol because stress also triggers the hypothalamus–hypophysis–suprarenal cortex axis via the PNEI interactive system. If the stressor remains for a longer time, mainly only increased cortisol levels persist.

### *The Reaction Stage*

In the reaction phase, the organism is trying to deal with the stressor. Three strategies are seen in mammals:

- **Flee like a horse.** The organism tries to avoid the new stressor by running away from it (e.g., by changing work or getting a divorce). This principle works very well for avoiding stress if flight is feasible. If it is not (e.g., if the individual is not qualified to change work or if no divorce is possible), then stage 3—exhaustion (see below)—will be entered over time.
- **Fight like a lion.** The organism fights back or even attacks the stressor (e.g., arguing with a boss or partner, exhibiting physical aggression, or having a dynamic response to the new situation). If a fight does not produce release of the stress, as time goes by and stress remains, then again the exhaustion stage will be reached.
- **Freeze like a donkey.** A third way of dealing with the stressor is a kind of pretending that the stressor is not there. The hope, by freezing, not to be detected by the stressor is a defense mechanism that is often seen in nature. Pheasants will hide from the hunter in the grass, ostriches will put their head in the sand, and donkeys and sheep will not move at all while being repeatedly attacked by a dog. In human contexts, we see people freezing in fear or psychologically hiding if, for example, a question is asked (head down and back round and leaning forward to look as small as possible, trying not to attract the attention of the stressor). If freezing does not stop the aggression of the stressor and does not lead to relief, once again in time a stage of exhaustion will also be reached.

### *The Exhaustion Stage*

An extended reaction stage will lead to exhaustion of reserves. In the exhaustion stage of the GAS model, most of the body's resources are depleted and the body is unable to maintain normal function. Because normal function of steering mechanisms is needed to keep ARSs running properly, the first physical symptoms of prolonged stress might appear. These symptoms are triggered by the autonomic nervous system (e.g., nausea, sweating, increased heart rate, and, via epinephrine and norepinephrine, eventually a high cortisol level with its consequences).

If the exhaustion stage is extended, long-term damage may result. The capacity of glands, especially the adrenal gland, and the immune system may become exhausted and function may become impaired, resulting in decompensation. The result can manifest itself in obvious illnesses, such as ulcers, depression, or even cardiovascular problems, along with other mental issues.

## Psychosomatism and Psychosomatic Diseases

Research has clarified the relationship between stress and psychosomatic diseases. The PNEI interactive system is the main cause for the somatization (especially immunologically), after extended periods of emotional stress, dysregulating bodily systems.

Although gastrointestinal ulceration, cardiovascular disorders, and migraine are mainly mentioned as psychosomatic diseases, research has also shown a relationship between stress and cancer, autoimmune diseases, and other severe diseases. Some authors even refer to the "social psychology of the immune system" to point to the

---

[54] Seyle H. A syndrome produced by diverse nocuous agents. *Nature.* 1936. [Reprinted in J Neuropsychiatr Clin Neurosci. Spring 1998;10(2):230–231.]

issue of stress and emotions influencing the immune system[55].

There is a clear relationship between extended stress conditions that cause exhaustion and psychosomatic disorders and diseases (created via the PNEI system) that cause tissue damage.

### *The Neural Reflex System*

Excitation is the activity produced in a nerve cell as a result of stimulation. Nerve irritation is the stimulating impulse to excitation. Irritation and excitation on the one hand, and inhibition of nerve activity on the other, are well balanced in a healthy organism. Nerve cells will "fire" in repetition in small groups and then stop "firing" their signals. The subtle balance can be disturbed in pathological conditions, as seen in inflammation processes.

In immune reactions, the neural reflex system takes part in the greater defense system. Nerve irritation and excitation are characteristic of the neural reflex system. For example, nerve excitation will intervene in body temperature regulation, causing shaking and muscle contractions in cold or feverish conditions. Via the PNEI system, it will intervene in hormone regulation, directing inflammation.

The neural reflex system enables steering shortcuts in correcting dysregulations in ARSs, which may need time to correct deviations that for some conditions might not be biologically purposeful. Neural reflex regulation appears in arterial pressure correction in pathological conditions[56]. The neural reflex system is directive and fast acting.

[55] Kaplan HB. Social psychology of the immune system: a conceptual framework and review of the literature. *Soc Sci Med.* 1991;33:909–923

[56] Vasquez EC et al. Neural reflex regulation of arterial pressure in pathophysiological conditions: interplay among the baroreflex, the cardiopulmonary reflexes and the chemoreflex. *Braz J Med Biol Res.* 1997;30 (4):521–532

### *Detoxification in the Liver*

Although the detoxifying function of the liver is discussed more in depth in Chapter 4, we here briefly discuss the liver within the greater defense system. The liver is one of the most important organs in the body for detoxification. Its function lies mostly in biotransformation; either making fat soluble toxins more water soluble for excretion in the kidneys or direct excretion in the bile, and biotransforming other toxins to make them less harmful.

#### *Regulation, Synthesis, and Secretion*

Liver cells (hepatocytes) have metabolic functions. They take up glucose, minerals, and vitamins from the blood and store them. Hepatocytes also produce various substances needed by the body, such as blood clotting factors, transporter proteins, cholesterol, and bile components.

The liver plays an important role in maintaining homeostasis; it regulates blood levels of cholesterol and glucose:

- Glucose is stored or released by the liver as needed. It reacts to two steering hormones from the pancreas: insulin and glucagon.
- Most blood proteins (e.g., albumin) are synthesized and secreted by the liver; decreased amounts of serum albumin may lead to *edema* (swelling due to fluid accumulation in the tissues). The liver also produces most of the proteins responsible for blood clotting, (*clotting factors*).
- Bile is synthesized by hepatocytes. The liquid is secreted into the bile duct and stored in the gallbladder. When it is needed metabolically (for digestion), it is released into the duodenum.
- Cholesterol is an essential component of cell membranes. The liver synthesizes cholesterol, which then circulates in the body to be used or excreted into bile for removal. Increased cholesterol concentrations in bile may lead to gallstone formation. The liver also synthesizes lipoproteins, which circulate in the blood and shuttle cholesterol and fatty acids between the liver and body tissues.

*Storage*

The liver stores glucose, in the form of glycogen, fat-soluble vitamins (A, D, E, and K), vitamins $B_6$ and $B_{12}$, and minerals such as copper and iron. However, excessive accumulation of certain substances can be harmful. When there is a lack of glucose in the blood, the liver transforms glycogen back to glucose, which is then released into the blood.

*Purification, Transformation, and Clearance*

The liver is the main detoxification organ in the body. It removes harmful substances from the blood and breaks them down into less harmful compounds. It also converts most hormones and drugs to their metabolites. The liver also transforms fat-soluble toxins into water-soluble ones, which are then excreted in urine via the kidneys and in bile.

- The liver converts ammonia to urea, which is excreted in urine by the kidneys. This process is called *deamination*. The liver can also convert one amino acid into another (but not the eight "essential" amino acids), a process called *transamination*.
- Bilirubin, a breakdown product of the red blood cell hemoglobin, is eliminated from the blood by the liver and is excreted in the bile.
- The liver plays a main role in hormonal modification and inactivation (e.g., the steroids testosterone and estrogen are metabolized by the liver). Thus, the liver plays an important role in hormone homeostasis.
- Nearly all drugs pass through the liver, where they are modified or biotransformed, becoming harmless waste products.
- The liver is generally responsible for detoxifying most chemical agents and toxins in the body.

**Figure 1.20** shows the important role of the liver in detoxifying the body.

*Fighting Infections*

The liver plays a vital role in fighting infections, particularly infections arising in the bowel. It does this by mobilizing part of the body's defense mechanism called the macrophage system. Half of

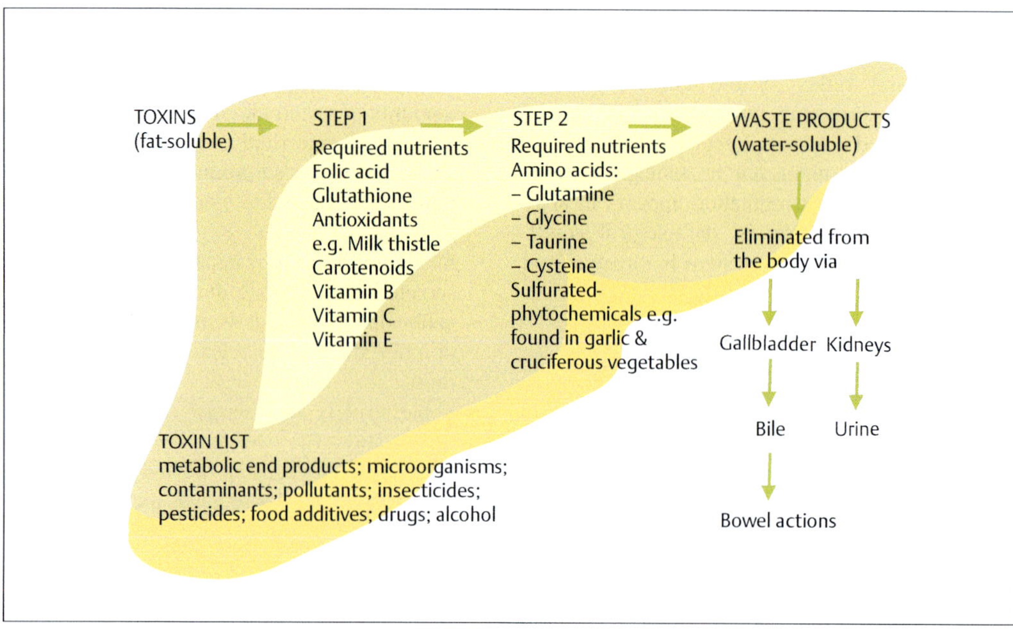


**Fig. 1.20** The liver: detoxification pathways.

the body's supply of macrophages (Kupffer cells) is found in the liver. Bacteria are attacked and destroyed by this macrophage system.

## The Matrix and the Mucosal Membranes

We have discussed the function of the matrix (ECM) earlier in terms of "The Ground Regulation System of Alfred Pischinger" and will discuss it further in Chapter 4. The function of the mucosal membranes is also discussed in detail in Chapter 4 in terms of drainage and detoxification.

## The Merging of the Six Subsystems of the Greater Defense System to Reach a Synergetic Concept

All six subsystems of the greater defense system work together in protecting the organism from antigens and toxins. The synergetic end effect means that toxins, metabolic waste products, and antigens are not only captured but are also transformed to nontoxic end products and eliminated out of the body as soon as possible.

The greater defense system, with all its subsystems, is essential for the survival of the organism and will often be the terrain of therapeutic action in bioregulatory medicine.

# 2 Disease Development

A well-defined pathological condition—a disease—is often seen as a static phenomenon. In reality, diseases are dynamic, changing, even evolving, and exhibiting different stages over time.

The same toxins, antigens, metabolic waste products, and other negative stimuli that we can characterize as homotoxins may produce different symptoms in the some organism at different stages of the evolution of the disease. In the homotoxicological view, the homotoxins are the real cause of the disease, disturbing normal cell function and causing deregulation of homeostasis.

## The Term Homotoxin

The concept of a homotoxin derives from the homotoxicological paradigm. The term homotoxin denotes a substance that has a disruptive effect on an organism as a whole. Generally speaking, we differentiate between exogenous and endogenous homotoxins.

As Paracelsus said, in addition to the nature of the substance in question, the dosage is the deciding factor. The physical condition of the patient also plays a decisive role, as does the extent to which the same intoxication is repeated (and possibly causes a cumulative effect).

> ***Exogenous homotoxins*** *are toxins that are toxic before entering the body (e.g., bacteria, viruses, and heavy metals).*
> ***Endogenous homotoxins*** *are substances that are produced by the body itself during metabolic processes and that are not metabolized properly and or excreted. They become toxic in the body itself (e.g.,* $CO_2$*, lactate, and calcium oxalate deposits).*

## The Homotoxicological Definition of Disease

From a homotoxicological viewpoint, illness is caused by the body's reaction to the presence of disruptive homotoxins. The recognized clinical symptoms of illness are what surface after the autoregulating system (ARS) has reacted to the threat. This means that illness is not the presence of symptoms as such, because these should only be seen as proof of ongoing defense activity.

As long as clinical symptoms are only viewed as a threat to the quality of life of the patient and the entire treatment is geared to the removal of these symptoms, the results will be superficial and the patient's long-term health is actually being compromised.

A biotherapeutic regulatory treatment considers the causative homotoxins and, by stimulating the body's own ARSs, will affect the actual causes of the illness. Biotherapy is always a regulatory therapy and never a suppression therapy.

> *Diseases are the* ***expression*** *of* ***biologically purposeful defense mechanisms*** *against endogenic and exogenic homotoxins or the expression of the organism's efforts to compensate for toxic damage it has sustained.*

The following will take a closer look at this definition and interpret the individual terms in the phrase "expression of biologically purposeful defense mechanisms."

**Expression:** What we see is not what we get. The symptoms are only the result of a defense activity of the organism against disruptive influences. Causal treatment means that we will have to do something about the disrupting homotoxins, which can be done by the regulation of the defense activity. Mere suppression of the symptoms is comparable with pushing an iceberg under the water, hoping that it will never reappear. If the pushing stops, the iceberg reappear. This phenomenon explains recurrence in classical medicine.

Another comparison can be made. Disease symptoms are only an expression of something deeper, just as the words a person uses are only an expression of his or her thoughts. Suppression of words, a ban on speaking, can never change the cause of talking, which is in the mind of the speaker. Treatment of the mind, as in psychotherapy, will automatically result in different expressions. Treatment of the mind is more effective in the long-term because it is a causal treatment. In the same way, the suppression of fever in viral infections will seem to be effective in the short-term. In the long-term, it will only increase the proliferation of the virus because fever has a microbicidal effect.

**Biologically:** *Bios* means "life" and *logos* means "word" or "study"; thus *biological* means "in accordance with life."

**Purposeful:** The concept implied by this term is extremely important in the homotoxicological definition of disease. It means that the reaction of the defense system will be in proportion to the need to reach the target, which is the elimination of homotoxins. Every regulatory aspect that homotoxicology is encompasses is incorporated in this concept of purposefulness. For example, autoimmune diseases are an inappropriate reaction of the defense system, as is hay fever. The reaction of the defense system is not in appropriate relationship to the danger posed by the aggressor (dust or the pollen).

**Defense mechanism:** Is part of a system that will react according to defined rules. The regulatory aspect is important: blockage of these regulatory mechanisms compromises the system and there is a risk of chronic involvement[1].

[1] Van Brandt B. *Inflammation Means Healing*. 3rd ed. Aartselaar: Inspiration Publishing; 2004

## *The Disease Evolution Table (DET) as a Practical Instrument in Daily Practice*

The disease evolution table (DET) is a two-dimensional graphic representation of the progression or regression of illnesses in six successive phases in response to how the body's defense system reacts to the presence of homotoxins (exogenous and/or endogenous) (see **Fig. 2.1**). Along the horizontal axis are the phases of diseases in order of severity. The horizontal axis also represents how the body deals with the homotoxins present. On the vertical axis are represented the different tissues/organs according to their embryological origin. In principle, each presenting clinical condition or disease state can be classified within this six-phase DET.

In consulting the DET, it is important to remember that the body will activate its natural defense mechanisms at three major pathophysiological levels or phases (humoral, matrix, or cellular) in an attempt to inactivate or detoxify and eliminate the homotoxins at each level. Each of these major pathophysiological phases is further subdivided into two other phases. If not successful at the first stage, the body will attempt again in the next phase. The illnesses will become more serious and the attempts to treat them will become increasingly difficult as the phases progress toward the last phases on the right of the table.

The DET is a very useful conceptual and practical means for clinically evaluating and following the natural tendency or evolution of the biological disease processes in the patient. To be able to classify the current status or stage of a patient in such a table, and to be able to follow the patient's clinical history, it is necessary to become familiar with the interpretation of each axis in the DET. Depending on the patient's location on the DET, specific antihomotoxic therapeutic strategies will need to be instituted to induce positive curative changes.

Thus, not only does the table categorize the stage and severity of pathological processes, it also becomes the basis for determining or planning the best therapeutic approach to ameliorate the patient's health status.

A full version of the DET, as a working instrument, appears on pages 38 and 39 as well as on the poster at the back of the book.

### *The Embryonic Differentiation of Tissues (DET Vertical Axis)*

Tissues are classified vertically according to their development from the three embryonic layers (ectodermal, endodermal, mesodermal).

### *The Six Phases of Disease Evolution (DET Horizontal Axis)*

The horizontal axis of the DET shows six successive phases in which diseases may occur. Patients do not have to pass through all the phases individually and successively to reach a phase at the right side of the table. The patient's history may show a zigzag pattern, or some phases may be skipped and will not occur at all in the history.

In fact, the six successive phases refer to the characteristics of the relationship between the organism and the presence of homotoxins and how the organism reacts to their disturbing presence (with regulation or compensation mechanisms). Initially, we see how the body reacts to intoxication as a trial for regulation and repair of homeostasis. Further on, we see what the intoxication does to the body, pushing it to compensation strategies.

#### *The Humoral Phases*

The humoral phases on the DET comprise the excretion phase and the inflammation phase, which are related by their "humoral" characteristics. Both phases are characterized by the final excretion of homotoxins, via direct excretion pathways or a complex process of inflammation. Excretion and inflammation are both strategies of regulation.

**Excretion Phase**

The excretion phase covers all the hypersecretions (endocrine) and hyperexcretions of the body in different organs and tissues. Because those secretions and excretions are increased compared with the normal standards in the population, they may be seen as a first stage of disease. Of course, the presence of homotoxins is a dormant danger and elimination and detoxification are needed. However, under normal conditions, detoxifying organs and excretion systems will eliminate them without manifestation of any significant clinical signs and symptoms because this is merely an amplification of a physiological process.

In this case, although there is a certain level of intoxication, in the normal course of living the body deals with it almost passively without causing any significant clinical manifestations typical of defensive reactions. Thus, the elimination of toxins occurs as a normal increased excretion process and the patient has no other clinical complaints.

**Inflammation Phase**

Once homotoxins reach the extracellular and/or intracellular space, the body will begin some form of a local defense reaction to counter the "intoxication" status. The appearance of this local "inflammatory" reaction is the reason why the patient is considered to be in an "inflammation phase." All acute inflammations are classified within this phase and should be seen as regulatory cleansing mechanisms.

Indeed, this first inflammatory reaction should be seen as a welcome natural and physiological attempt by the organism to eliminate the toxins. The activation of phagocytes and the process of phagocytosis should be seen as the first reactive step of detoxification. All of the classic characteristics of inflammation might be present: swelling, redness, pain, temperature increase, and loss of function in the affected tissue.

Acute inflammation should thus be seen as a "cleaning" process of the matrix as the ultimate goal is to get rid of the homotoxins. The cell per se is not yet involved, although the inflammatory processes can passively damage the cell (e.g., when free radicals are released by "frustrated" or "overzealous" neutrophils). This inflammation is hallmarked by normal angiogenesis and mostly complete tissue restitution

HEALTH → Status of Regulation / Deregulation → DISEASE

| Organ System/Tissue | | Humoral Phases: Excretion Phase | Humoral Phases: Inflammation Phase | Matrix Phases: Deposition Phase | REGULATION/COMPENSATION DIVISION | Matrix Phases: Impregnation Phase | Cellular Phases: Degeneration Phase | Cellular Phases: Dedifferentiation Phase |
|---|---|---|---|---|---|---|---|---|
| ECTODERMAL | 1. EPIDERMAL | Increased Sweating, Cerumen, Sebum, Smegma | Dermatitis, Impetigo, Abscess, Furuncle, Otitis externa | Hyperkeratosis, Seborrhoic eczema, Naevus, Skin Tags (soft wart) | | Atopic eczema, Urticaria, Warts, Fissura ani, Acne rosacea, Hirsutism | Psoriasis, Decubitus ulceration, Radiation injury, Pemphigus vulgaris | Squamous cell carcinoma, Basal cell carcinoma, Melanoma |
| ECTODERMAL | 2. ORODERMAL | Hypersalivation, Hyperlacrimation | Otitis media, Pharyngitis, Stomatitis, Gingivitis, Apthous ulceration, Glossitis, Rhinitis (acute), Sinusitis (acute), Laryngitis, Dental abscess | Nasal polyp, Eustachian tube catarrh (serous otitis media), Dental granuloma | | Atopic Rhinitis, Hay fever, Sinusitis (chronic), Rhinitis (iatrogenic), Anosmia, Menière's syndrome, Hypoacusis | Otosclerosis, Deafness (transmission), Ozaena, Atrophic rhinitis, Dental caries, Parodontosis | Leucoplakia (orodermal), Cancer of the tongue, Laryngeal cancer, Nasopharyngeal cancer, Tracheal cancer |
| ECTODERMAL | 3. NEURODERMAL PNS and CNS | Increased secretion of neurotransmitters | Neuralgia, Neuritis, Polyneuritis, Meningitis, Encephalitis, Trigeminal neuralgia (acute) | Neuroma, Amyloid deposition, Heavy metal deposition | | Epilepsy (petit mal), Paresis, Tics, Neuritis (toxic), Attention-Deficit/Hyperactivity Syndrome (ADHS), Guillain Barré syndrome, Poliomyelitis (acute), Trigeminal neuralgia (chronic) | Parkinson disease, Epilepsy (grand mal), Alzheimer's disease, Multiple Sclerosis, Amyotrophic Lateral Sclerosis, Peripheral Neural Atrophy, Diabetic neuropathy, Neurofibromatosis | Glioma, Meningioma, Astrocytoma |
| ECTODERMAL | 4. EYE | | Conjunctivitis (acute) | Pterygium, Mouche volante (floaters), Iris spot (initial) | | Uveitis, Allergic conjunctivitis, Iris spots (chronic) Iritis, Astigmatism, Myopia, Presbyopia, Keratoconus, Pannus, Arch (senile) | Glaucoma, Cataract, Hemianopsia, Macular degeneration, Paralytic mydriasis | Retinal cancer, Retinoblastoma |
| ECTODERMAL | 5. SYMPATHICODERMAL | Increased adrenalin and noradrenaline secretion | Flushes, Hypervagotony, Hypersympathicotonus | Ganglion neuroma | | Dysautonomia (including Orthostatic hypotension) | Addison's disease, Reflex sympathetic dystrophy (RSD) or (Sudek's syndrome), Horner's syndrome | Pheochromocytoma, Neuroblastoma |
| ENDODERMAL – 6. MUCODERMAL | 1. Respiratory | Sputum | Bronchitis (acute), Tracheitis | Nasal polyp | | Bronchitis (asthmatic), Chronic tracheitis (viral), Cystic fibrosis | COPD (chronic obstructive pulmonary disease), Atrophy of bronchial mucosa | Tracheal cancer, Bronchial cancer |
| ENDODERMAL – 6. MUCODERMAL | 2. Digestive | Increased digestive juices | Oesophagitis (acute), Gastritis (acute), Gastroenteritis (acute), Colitis | Gastric polyps, Intestinal polyps, Obstipation, Melanosis of the colon | | Gastric ulcer, Duodenal ulcer, Gluten enteropathy(mild), Leaky Gut Syndrome, Dysbiosis | Crohn´s disease, Colitis ulcerosa, Atrophy of the small intestinal villi, Gluten enteropathy (severe) | Barret's esophagus, Esophageal cancer, Gastric cancer, Duodenal cancer, Rectal cancer |
| ENDODERMAL – 6. MUCODERMAL | 3. Urogenital | Increased mucous production | Bartholinitis, Cystitis, Urethritis, Infections of the urogenital mucosa | Bladder polyps, Uterine polyps | | Interstitial cystitis | Atrophy of the urogenital mucosa | Bladder cancer, Cervical carcinoma |
| ENDODERMAL – 7. ORGANODERMAL | 1. Exocrine Sexual | Lactorrhoea | Mastitis | Mammary cysts, Breast calcifications | | Mammary fibroadenoma, Fibrocystic mastopathy | Breast atrophy, Gynecomastia | Mammary carcinoma |
| ENDODERMAL – 7. ORGANODERMAL | 2. Exocrine Digestive | Increased bile salt secretion, Increase gastric acid secretion | Pancreatitis, Sialitis | Cholelithiasis, Steatosis hepatica, Pancreatic calcifications, Pancreatic cysts, Liver cysts, Wilson's disease, Salivary gland calcifications | | Chronic hepatitis, Chronic pancreatitis, Viral pancreatitis (e.g. Mumps), Alcoholic hepatitis, Cystic fibrosis | Hepatic cirrhosis, Hepatic Iatrogenic disease | Liver cancer, Pancreatic cancer |
| ENDODERMAL – 7. ORGANODERMAL | 3. Respiratory | | Acute pulmonary abscess, Pneumonia | Bronchiectasis, Pneumoconiosis | | Bronchial asthma, Cystic fibrosis | Emphysema, Chronic pulmonary abscess, Interstitial fibrosis of the lung, Fungal balls | Pulmonary cancer |
| ENDODERMAL – 7. ORGANODERMAL | 4. Endocrine | Increased thyroid hormones, Parathyroid hormones, Thymic hormones, Insulin, Glucagon, Enteric hormones, Cortico suprarenal hormones, Adeno hypophyseal hormones | Thyroiditis , e.g. de Quervain's thyroiditis | Thyroid cysts, Adrenal cysts, Adrenal adenoma Hypophyseal adenoma , Thymoma, Insulinoma, Parathyroid gland adenoma, Thyroid goiter, Adrenal adenomas | | Grave's disease, Hashimoto's disease (1st stage), Puerpural thyroiditis, Cushing's syndrome, Precocious puberty, Adrenal exhaustion | Hashimoto's disease (2nd stage), Riedel's thyroiditis, Parathyroid atrophy | Thyroid cancer, Parathyroid cancer, Adrenal cancer, Carcinoid syndrome |

| MESODERMAL | | | | | REGULATION/COMPENSATION DIVISION | | |
|---|---|---|---|---|---|---|---|
| MESENCHYMAL | 8. CONNECT. TISSUE | Increased secretion of metalloproteinases, Increase in glycoprotein formation | Abscess, Reactive inflammatory response of the matrix, Tendinitis | Lipoma, Storage of toxins in the matrix, Amyloidosis, Mucopolysaccharidosis, Periarthritis humeroscapularis calcinosa | Mixed connective tissue disease (MCTD), Marfan's syndrome, Ehlers-Danlos Syndrome, Sphingolipidosis | Scleroderma, Carbohydrate deficient glycoprotein syndrome, Peyronie's disease, Progeria, Dupuytren's contracture | Sarcoma |
| MESENCHYMAL | 9. OSTEODERMAL | | Osteomyelitis, Chondroitis | Osteophyte formation, Bone cysts | Osteomalacia, Early osteoporosis | Osteoporosis, Paget's disease | Osteosarcoma |
| MESENCHYMAL | 10. HEMODERMAL – 1. Blood | | Leukocytosis neutrophila, Anaemia related to acute infection | Thrombocytosis, Polycytemia (reactive), Hypercoagulation | Eosinophilia, Leukopenia, Anaemia (including anaemia of chronic disorders), Hypercoagulation | Aplastic anaemia, Thrombocytopenia, (including Idiopathic thrombocyopaenic purpura), Pancytopenia, Vaquez's disease | Leukemia |
| MESENCHYMAL | 10. HEMODERMAL – 2. Heart | Increased cardiac output, Tachycardia | Myocarditis, Extrasystoles, Acute rheumatic fever | Left ventricular hypertrophy, Coronary atheroma | Angina pectoris, Atrial enlargement, Arrhytmia cordis, Rheumatic fever affecting the heart, Prolapse of the mitral valve (Barlow's syndrome), Cardiomyopathy | Myocardial Infarct, Ventricular arrhytmia, Stenosis and insufficiency of the cardiac valves | Sarcoma |
| MESENCHYMAL | 10. HEMODERMAL – 3. Vascular | Increased production of endothelial mediators | Phlebitis, Arteritis, Endothelial inflammation | Venous stasis, Arterial plaques (atheroma), Haemorrhoids | Vasculitis, Arteriosclerosis, Varicose veins, Panarteritis nodosa, Angioma, Varicocele | Peripheral vascular disease, Aneurysm, Arteritis obliterans, Peripheral vascular disease | Angiosarcoma |
| MESENCHYMAL | 11. LYMPHODERMAL | Increased lymph production | Tonsillitis, Adenitis, Adenoiditis, Lymphangitis | Lymph edema, Lymph adenopathy, Tonsillar hypertrophy, Adenoid hypertrophy | Indurated edema, Venerial lymphogranuloma, Cat scratch disease. | Lymphatic tuberculosis, Elephantiasis | Lymphoma (Hodgkin's, Non Hodgkin's), Lymphosarcoma |
| MESENCHYMAL | 12. CAVODERMAL | Increased synovial liquid, Cerebrospinal fluid | Arthritis, Polyarthritis, Synovitis, Acute rheumatic disease | Hydrops (articular), Gouty tophi, Haemarthrosis | Chronic arthritis, Reiter's syndrome, Hydrocephaly, Spinal disc herniation | Arthrosis, Ankylosing spondylitis | Sarcoma, Chondrosarcoma |
| | 13. NEPHRODERMAL | Frequent urination | Nephritis, Glomerulonephritis, Pyelitis | Nephrolithiasis, Renal cysts, Renal sand, Orthostatic albuminuria, Haematuria | Pre-clinical nephrosis, Nephrotic syndrome, Chronic hematuria, Goodpasture's syndrome, Auto-immune glomerulonephritis | Nephrosis, Chronic glomerulonephritis, Tuberculosis of the urogenital tract | Hypernephroma, Wilms' tumor, |
| | 14. SERODERMAL | Increased production of serous fluid | Pleuritis, Peritonitis, Pericarditis | Pleural effusion | Chronic exsudative pleuritis and serositis, Ascites, Chronic pericarditis | Pleural, Pericardial and peritoneal tuberculosis, Pleural adhesions | Mesothelioma, Primary peritoneal carcinoma, Primary pleural cancer |
| | 15. GERMINODERMAL | M Increased seminal fluid | Prostatitis, Epididymitis, Orchitis | Spermatocoele, Early benign prostatic hyperplasia (BPH) | Benign prostatic hyperplasia (BPH), Oligo asthenospermia | Sterility | Prostate cancer, Testicular cancer, Seminoma, Teratoma |
| | 15. GERMINODERMAL | F Heavy menstruation | Ovaritis, Adnexitis, Metritis, Dysmenorrhea | Ovarian cysts, Uterine polyps, Uterine Fibroids | Chronic adnexitis, Amenorrhea | Infertility, Ovarian atrophy | Ovarial cancer, Ovarial teratoma |
| | 16. MUSCULODERMAL | Myalgia | Myositis | Myogelosis, Myositis ossificans | Muscular asthenia, Mitochondrial myopathy, Autoimmune dermatomyositis | Muscular atrophy, Muscular dystrophy | Myosarcoma |
| | | Self regulation. Self-healing effects. Favourable Prognosis. | | | Compensation. Tendency to aggravation. Doubtful Prognosis. | | |

**Fig. 2.1** The disease evolution table (DET). **Note:** This table can also be found on a poster at the back of this book.

*The Matrix Phases*

The matrix phases comprise both the deposition and impregnation phases because the ECM is central in both.

**Deposition phase**

This phase is an expression of the body's incapacity to eliminate (excrete) homotoxins, when predominant storage of toxins within the ECM begins to cause disorders of regulation. It is reached when the body has to temporarily store (deposit) toxins. This may occur for two reasons: (1) the inflammatory process (the previous phase) was not adequately activated or was blocked or suppressed (e.g., by anti-inflammatory drugs); or (2) the excretion mechanisms are hypofunctional or the toxic load is excessive. Consequently, if inflammation pathways are blocked or the amount of homotoxins gets out of hand, the organism will choose a process of (temporary) storage or deposition of the homotoxins.

Clinically, this phase is a relatively silent process with very few clinical signs and symptoms; however, it is a quite dangerous process. It is only a question of time before the homotoxins will infiltrate into the cell or from outside the cell will interfere with and have many effects on normal cellular functions.

**Impregnation Phase**

Once the homotoxins begin to "impregnate" the ECM or infiltrate within the cells, inducing intracellular effects, diseases of the impregnation phases appear. Homotoxins practically become part of the structural components of the connective tissue and the matrix. Some toxins (e.g., viruses) may also directly penetrate cells within the connective tissues and/or cells of the parenchyma. Toxins that reach this stage will begin to induce functional changes both in the matrix and in cells, such as blocking of enzymes or metabolic pathways and compromise of the mitochondrial respiratory chain. We see less efficient functioning of the cell, and the reactions of the organism toward the homotoxins are often not purposeful; a minimal load of specific homotoxins produces an overreaction of the organism's defense mechanisms (e.g., asthma, hay fever, migraine, or gastric ulcer). Histologically, some changes in structural components begin to become evident. Clinically, signs and symptoms that are indicative of cellular damage will appear.

A number of diseases classified in this phase will manifest as low-grade chronic inflammation with episodes of acute exacerbation, such as asthma or inflammatory bowel disease.

The impregnation phase can be reached in a very short time, depending on the characteristics of the homotoxins. Most viruses will attempt to enter a host-cell and proliferate rapidly; although the organism will try to develop a specific defense (immunoglobulins) and eliminate the infected cells (T-cell activity and NK cell–induced elimination), the immediate situation is an impregnation phase because of the intracellular presence of the homotoxins (the virus). Sometimes, even if tissues are fully restored and the lost cells are replaced, the viral condition remains in the impregnation phase while the virus is present if the virus is incorporated into the genetic material of the host cell. In postviral syndromes, this situation can last for a long time, even years.

*The Cellular Phases*

In the cellular phases, the cell itself is involved and the diseases are purely cell related. The cellular phases comprise the degeneration phase and the dedifferentiation phase. These phases are hallmarked by chronic inflammation, tissue scarring, and also abnormal angiogenesis, which is seen in degenerative diseases such as osteoarthritis as well as in cancerous processes. The organism will try to compensate the damage that occurs in certain tissues. Regulation mechanisms fail.

**Degeneration Phase**

The natural defense system is no longer able to eliminate or excrete toxins from the cells and/or the matrix, or the imbedded homotoxins cause disruption of the cell-to-cell communication as well as the matrix-to-cell communication. Intracellular structures, including genetic components, cellular membranes, and even extracellular groups and systems of cells, become increasingly and seriously damaged. Predominantly cellular damage occurs in this phase. The progressive intoxication causes complete functional loss of the affected cells until they die. In the long term, we see tissue loss and limited function of the whole affected tissue. By definition, the degeneration phase is involved in chronic degenerative diseases, most of them irreversible.

**Dedifferentiation Phase**

The dedifferentiation phase is involved in all diseases in which abnormal cell proliferation (tissue growth) is the main characteristic. Cells lose their specificity and dedifferentiate to totipotent cells (inverted embryological specificity) that can easily also lose their restraining controls and begin to migrate to other locations in the body (metastases). In this phase of complete degeneration, the body becomes increasingly influenced by endogenous homotoxins (i.e., toxins generated within the body through cellular destruction). All malignant tumors (cancers) are classified here.

### *The Regulation/Compensation Division*

Between the third (deposition) and fourth (impregnation) phases, there is the so-called regulation/compensation division (RCD). This is an artificial dividing line that demarcates the point that separates the ability of a patient to self-regulate or merely to compensate for the damage if there is no external help.

This boundary is very important physiologically, pathologically, and clinically—and thus therapeutically—because it is the separating line between (i) the phases in which simple excretion of toxins is still possible and tissue restoration is achieved, and (ii) the impregnation phase, in which simple excretion is no longer achievable, at least spontaneously, and the healing and compensation are characterized by destruction of tissues (fibrosis or metastatic cancer growth).

To the left of the RCD, the tendency of the organism is to regain homeostasis. Deregulation is corrected, and intoxications are purposefully drained and detoxified if possible. To the right of the RCD, the organism will change its strategy and choose compensating measures to keep it functioning for as long as possible. Because regulatory capacities are more or less lost, the organism seeks compensation for the losses.

Clinically, the prognosis is quite good to the left of the RCD. Pathological conditions react very well using simple therapy schemes in bioregulatory medicine as ARSs remain intact and can be influenced. To the right of the RCD, the prognosis is more dubious and there is a tendency for aggravation. Bioregulatory medicine can be very effective, but the therapy schemes will be more complex because of the need for more pillars of antihomotoxic treatment (see Chapter 4).

## *Classification on the Table*

Because disease is dynamic, it is useful to classify a patient on the DET each time the patient is seen in consultation. It is only then that the evolution of the patient's condition can be correctly evaluated homotoxicologically. Relocations from left toward the right and from top toward the bottom on the table are signs of disease aggravation or *disease evolution*. The movements in the opposite directions are signs of disease amelioration or *health evolution*. Thus, the patient's history is essential for a good understanding of what evolution is actually taking place and for determining the correct medications to support evolution toward a more healthy state (health evolution) or to counter a negative trend (disease evolution).

## *Decision Tree for Classification*

To classify a patient's current disease on the DET, we look first at the affected tissue. The vertical axis gives a classification of tissue going from more structured to less structured, such as the mesenchyme. This classification from the top toward the bottom of the table follows the sequence of embryonic development in four poles. Icons on the left of the DET will help recognize the affected tissue in the embryonic layers (ectodermal, endodermal, and mesodermal).

Secondly, the phase of the disease should be identified. The main characteristics of each phase should be compared with the clinical characteristics of the presenting disease. The main phase characteristics are as follows.

- **Excretion phase:** Hyperexcretions or hypersecretions are present. The excretion and secretion pathways are in overdrive, excreting or secreting many more substances than in "normal" conditions (compared to set point values). However, very few or even no clinical complaints are present.
- **Inflammation phase:** Acute inflammation is present and is confirmed clinically (symptoms including pain, swelling, redness, warmth, and loss of function of the affected tissue). The inflammation phase includes acute inflammation reactions only, not the chronic ones or those related to allergies. As already mentioned, the process is finite, with normal restitution of tissues and normal angiogenesis.
  Chronic inflammatory reactions may lead to cellular damage and are found to the right of the RCD. Allergies are repetitive inflammation reactions alternating with periods of latency, in which the patient is free of symptoms, and belong to the impregnation phases not the inflammation phases.
- **Deposition phase:** Storage of bodily liquids or substances from external or metabolic origin is present. The deposition of homotoxins does not need to show clinical complaints to be defined as a deposition phase. The fact of physical deposition is sufficient. Thus, even if polyps, cysts, and calculi give no complaints at all (at the moment), they should be classified as deposition phases.
- **Impregnation phase:** Diseases are in this phase when the effects of the intoxication are causing intracellular damage. Clinically, we may see acute exacerbation peaks of the disease alternating with periods of latency, often characterized by greater or lesser absence of clinical complaints. Many diseases, such as hay fever, bronchial asthma, migraine, gastric ulcer, and angina pectoris, are impregnation phases. Also, viral infections often begin with an impregnation phase because the virus infects the cell intracellularly. The acute inflammation seen in the evolution of a viral infection should be interpreted as a health evolution in which the patient evolves from an impregnation to an acute inflammation phase. The inflammation is an immediate attempt of the organism to get rid of the virus.
- **Degeneration phase:** In this phase diseases are mainly characterized by cell death. Most chronic degenerative diseases are classified here. The healing is through fibrosis and there is abnormal angiogenesis.
- **Dedifferentiation phase:** In diseases in this phase cells *dedifferentiate* (i.e., they partly lose their specificity and become totipotent). All tumors are classified here.

### *Consequences for Therapy*

Therapy schemes for a specific disease related to a certain position on the DET will have to conform to certain standards in antihomotoxic therapy. These standards are described further in Chapter 4.

The therapeutic approach differs according to the phase the patient is in. The types of antihomotoxic medications used will vary accordingly. To the left of the RCD, drainage and detoxification combined with immunomodulation are emphasized; to the right of the RCD, cell and organ support are added to the therapy strategy.

The main point regarding the DET is that the farther right and the farther down the table the patient's condition is, the more complex is the required antihomotoxic therapy protocol.

## Disease Evolution

Progression (evolution) of the patient's pathological status on the DET toward more serious conditions is too often for iatrogenic reasons.

Progression of the patient's signs and symptoms from the left to the right of the table or from the top to the bottom of the table, or a combination of both, represents a deteriorating condition and is called *disease evolution* (see **Fig. 2.2**). The inhibition or suppression of natural biological defense mechanisms (e.g., by the indiscriminate use of anti-inflammatory drugs) is frequently responsible for the progressive involvement of the body in other pathological conditions.

When toxins are inhibited from being excreted, they will not only impregnate locally but will also

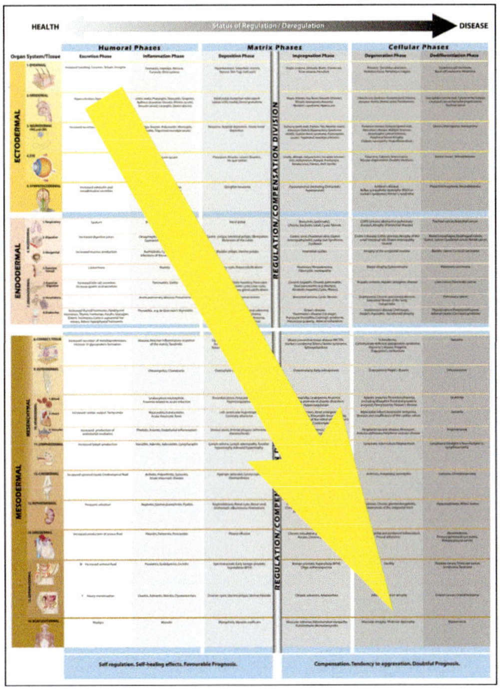


**Fig. 2.2** Disease evolution: the patient's pathological condition is worsening.

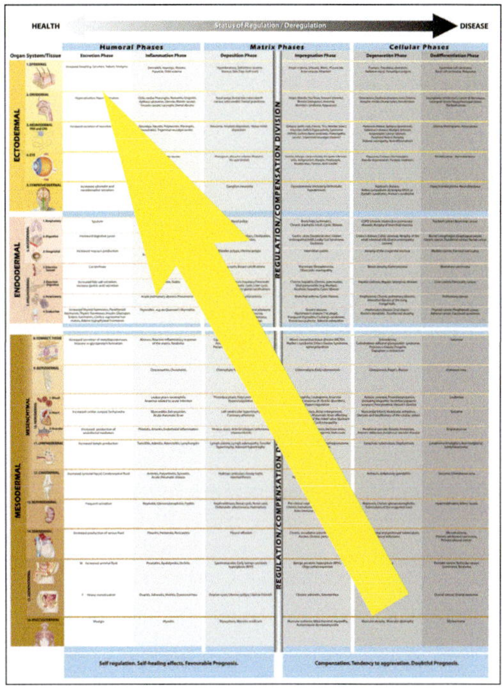


**Fig. 2.3** Health evolution: the patient's pathological condition is improving.

easily be transferred to other tissues, even at great distances from the original focal point:

- In **linear progression**, the toxins are transferred to other tissues of similar embryological origin (e.g., tonsillitis [lymphodermal]→polyarthritis [cavodermal]→ nephritis [nephrodermal]).
- In **disordered progression**, phases are skipped and tissues of different embryological origins are entered locally and/or at a distance.
- Here, the disease tends to evolve negatively in the ontogenetic and phylogenetic sense.

## Health Evolution

A regression in the patient's pathological condition with the relevant signs and symptoms (and actually with the reappearance of old signs and symptoms) from the right to the left of the table or from the bottom to the top of the table, or a combination of both, is an indication of an ameliorating process. This process is called *health evolution* (see **Fig. 2.3**).

In these cases, the disease tends to evolve positively in the ontogenetic and phylogenetic sense.

# 3 Preparation Groups in Antihomotoxic Medicine

Various different *preparation groups* are used within the range of antihomotoxic medications. The "architecture" of the medication's formula will change depending on the objective of the medication (i.e., the indication for which it is used). Although single preparation groups can be used in isolation, on many occasions antihomotoxic medications are composed of combinations of substances of different preparation groups, brought together to optimize the effect of the medication.

## Plant Preparation Groups (PPGs)

Plants are often sources of active components in medications. This is not only the case in bioregulatory medications, such as antihomotoxic agents, but also in conventional medications. Many examples exists: ASA (acetylsalicylic acid), a molecule used in nonsteroidal anti-inflammatory drugs (NSAIDs), originates from a tree (*Salix alba*); digitalin originates from a plant (*Digitalis officinalis*); taxine, used in a cancer medication (Taxol), is derived from *Taxus officinalis*; a plant extract from *Sabal serrulatum* is used in medications against prostate hypertrophy; and hypericine, an alkaloid from the plant hypericum, is used in the treatment of depression.

Plants or their components often have physiological effects that can be used therapeutically; the Arndt–Schulz rule (see Chapter 1) can be applied as minute dosages in medications to create the desired physiological effects.

The effects of plants in subtle dosages are described in a so-called *Materia Medica*, a book that describes in detail the physiological effects that plant extracts or their components have on humans. There is a strong correlation between what is described in toxicology books used in pharmacology and what is described in the *Material Medica*. The symptom pattern of intoxication will be found to be the same or similar in both reference sources; the inverse therapeutic dose–effect relationship—the goal of antihomotoxic medication—is exploited by giving the toxic substance to the patient in a minute dosage.

## Mineral Preparation Groups

Minerals are often used in conventional medications. In antihomotoxic medications, minute dosages of a mineral are given to counter the intoxication pattern that the patient shows.

## Organ Preparation Groups (Organ Regulators; ORGs)

Organ extracts have been used for many decades with the aim of strengthening the function and structure of organs. Because they contain complex proteins, organ extracts are also used as immunomodulators (see Chapter 4).

The principle of oral tolerance, in which minute dosages of specific proteins cause the formation of specific antigen-related $T_{reg}$ cells, down-regulating inflammatory processes, is behind the selective use of organ extracts. Inflammation is often antigen induced, and these antigens have characteristic proteins leading, via APC motif presentation, to the creation of antigen-specific proinflammatory $T_H1$ and $T_H2$ lymphocytes. Well-selected organ extracts, containing similar proteins to the inflammation-causing antigen, may induce $T_{reg}$ cell generation if given in minute dosages. TGF-β secreted by the $T_{reg}$ cells will inhibit the proinflammatory T-helper lymphocytes with a similar motif. This was postulated by Heine in 1998[1] and was confirmed in other research in 2002[2].

## Catalyst Preparation Groups (CPGs)

Essential catalysts of biochemical processes in the body, as found in the citric acid cycle, are often used in cell- and organ-supporting antihomotoxic medications.

Since Nobel Prize winner Sir Hans Adolf Krebs described the citric acid cycle as an essential and continuous metabolic process for providing the living cell with energy (cell oxygenation)[3], the understanding of its importance in fighting chronic degenerative diseases has gained momentum in antihomotoxic treatment. Catalysts, enzymes, coenzymes, and many other factors play an essential role in this complex mitochondrial process. The functioning of the cell is directly related to the efficient turnover of the Krebs cycle. The supply of factors to complete the chain depends on many variables. Nutrition is essential here, as is the transport of factors from the bloodstream to the cell and of waste products of the cycle back to the bloodstream.

The catalysts and coenzymes may be needed in minute dosages to enhance the mitochondrial process and ameliorate cell functioning. This is further discussed in Chapter 4, under Cell Support (p. 78).

[1] Heine H et al. Immunologische Beistandsreaktion durch pflanzliche Extrakte in antihomotoxischen Präparaten. *Biologische Medizin.* 1998;27(1):12–14

[2] Heine H et al. The anti-inflammatory action mechanism of an anti-homotoxic composite remedy. *Ärtzezeitschrift für Naturheilverfahren.* 2002;43:2

[3] Berg JM et al. *Biochemistry,* 5th ed. San Fransisco: WH Freeman; 2002

## Architecture of an Antihomotoxic Medication

The way an antihomotoxic formula is composed, its "architecture", is quite different from the simple architecture of a conventional medication (for which, in most cases, there is one active component) or the complex architecture of other compounded complementary medications found in bioregulatory medicine.

Antihomotoxic medications often conform to the Bürgi principle (see below). In addition to this principle, the components are chosen to be complementary and synergetic in their action against the indication.

### The Bürgi Principle

The Bürgi principle is a basic but often forgotten principle in pharmacology that says:

> The effect of two substances that lead to the same change in function or remove the same symptoms adds up when they have the same, and amplify when they have different, pharmacological target points.

In 1910, Emil Bürgi, lecturer on pharmacology at the University of Berne in Switzerland, had shown that the combination of two different therapeutic substances or medications with equal or similar activities can, when their points of action are different, produce total effects that are greater than the simple additive sum of their individual effects.

In addition to the a very selective choice of the components in antihomotoxic medications, the Bürgi principle will add a supplementary effect to the composition. Pharmacological publications describe the therapeutic effect of individual substances. However, if these substances have the same physiological objective but different sites of action, their combined effect will be a physiological change that is much greater than anticipated from the linear summation of the effects of the isolated individual substances.

The Bürgi principle should not be confused with the well-known synergetic effect of com-

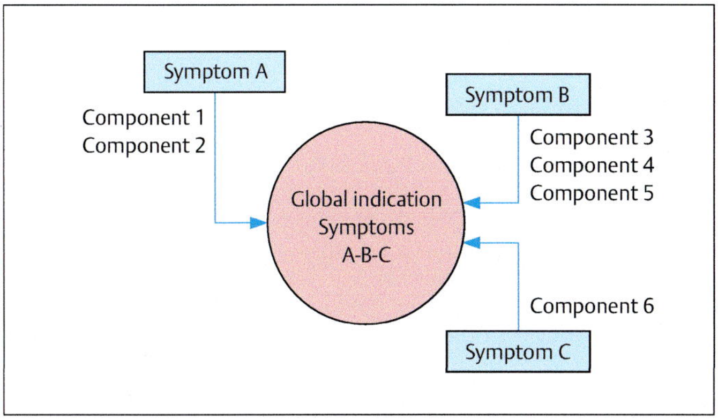


**Fig. 3.1** Different aspects of the global indication can be covered by different components of the formula.

pounds. In the Bürgi principle, the target of the compounds is the same but the site of action or interaction is different. In the synergetic effect, the global target is the same but the subtargets of each component are different.

### Complementarity

The physiological effects of compounds are well described in reference books. Each compound has one or more sites of action and also has a therapeutic physiological effect. This effect can be a *subeffect* of a global effect that the formula is seeking when complementarity comes into the architecture of an antihomotoxic drug.

Compounds in antihomotoxic medications often have this complementary effect (i.e., different components may cover different aspects or subsymptoms of the same disease or clinical indication). Thus, many different components that treat redness, swelling, and sensation of heat and pain, when brought together in one formula, treat inflammation as a global indication because inflammation as a process is characterized by these various subsymptoms.

The complementary and synergetic effects of the different components of a formula often go together. The synergetic effect is induced by the complementarity of the components used against a well-defined indication.

### The Synergy of Components

The synergetic effects of the components in antihomotoxic medications refer to the fact that the combined effect of the drug exceeds the individual effects of each component in the organism[4].

The synergetic effect of the different components in one formula can be compared with synergies we encounter in other complex cybernetic systems. Stimuli that trigger cybernetic systems often amplify each other so that the end result is greater than the summation of the effects of each isolated component (**Fig. 3.1**).

This synergetic effect is seen not only in the combination of different components in one formula in practice, but also in the combination of different formulas in one therapy scheme, as is often applied in biopuncture[5].

---

[4] Swayne J. *Dictionary of Homoeopathy.* Commissioned by European Commission Medicine Research Group; 1996

[5] Kersschot J. *Biopuncture in General Practice, A Practical Guide for Family Physicians.* Aartselaar: Inspiration; 2004

# 4 The Three Pillars of Antihomotoxic Treatment

Antihomotoxic treatment is one of the main players of the different types of bioregulatory medicine that intervene in cybernetic systems in the body. According to the concept of homotoxicology, the patient is evaluated in relation to his or her position on the DET. The patient's position on the DET and medical history will determine which of the three pillars of antihomotoxic treatment will be used in the final therapy protocol to treat the patient.

Antihomotoxic treatment is a world leader within bioregulatory medicine, as evidenced by the more than 90 million antihomotoxic injections administered worldwide by medical doctors every year and the millions of units of antihomotoxic medications applied in other galenic forms as adjuvant therapy.

To understand the strategy behind the successful application of homotoxicology, we need to discuss and study the three pillars of antihomotoxic treatment in detail.

Although most antihomotoxic medications have a clear and well-determined clinical indication for which they are used, more complex pathological conditions will need combination therapy of different types of antihomotoxic medications within a strategic framework to be successful in the long-term. The reason for this is that the farther right on the DET the pathology occurs, the more the cell itself is involved. There is an immense difference between extracellular and intracellular pathological conditions and there is a correspondingly immense difference in the appropriate antihomotoxic treatment strategies. We may state definitively that the farther right the patient's condition is located on the DET, the more pillars will be involved in treating the patient's condition effectively.

The three pillars of antihomotoxic treatment are:

1. Drainage and detoxification
2. Immunomodulation
3. Cell and organ support

Each pillar will be studied more in depth in the following

## Drainage and Detoxification

To better understand the rationale and methods for detoxification and drainage, it is useful to define what toxins are (their origin and nature), how they reach the body cells and tissues, and how they influence homeostasis and become etiological factors for inducing disease processes.

Any agent (physical, chemical, or microbial) that adversely modifies or damages a balanced biological system is considered a "toxin." Toxins may enter the body from the external environment (*exogenous*) (also called toxicants or xenobiotics) through the gastrointestinal system by ingestion, the respiratory system by inhalation, and the skin by passive absorption or injection. Toxins may also originate within the body itself (*endogenous*) as by-products of physiological metabolism (bilirubin, creatinine, or lactic acid) or as metabolites under abnormal metabolic conditions (excess production or degradation of neurotransmitters and/or hormones and excess free radical formation).

Modern medicine is quite successful in diagnosing and treating acute intoxications as medical emergencies for heavy metals and drug poisoning, as well as symptomatic subacute poisonings from a variety of toxins (chemicals, drugs, and/or other xenobiotics), but only if some laboratory evidence of intoxication is found. Unfortunately for the patient, subclinical or subtle states of chronic intoxications are not recognized, except that the appearance of any symptom is treated with suppressive medications independently of its etiology.

The biological effects of chronic subtle states of intoxication are of great importance to biological medicine. Paracelsus is often paraphrased as stating that "the dose makes the poison." Often, if the dose is not high enough to produce immediate acute effects and/or cannot be detected with our present technological limitations, then the toxin is seen as not responsible for the biological effects and influences on the health of the individual.

Another very important consideration to bear in mind is the individual's toleration of or susceptibility to specific toxins. The toleration of biological systems for a toxin is in part genetically determined and in part acquired on the basis of enzymatic induction and/or inhibition, the degree of functionality of the target organ, and the functional reserve capacity of specific organ systems. The clinical manifestations of biological effects of toxins depend on the physical and chemical properties of the toxin itself and also on the duration and route of exposure, the toxin's mechanism of action, and individual susceptibility.

Chemical compounds, which compose the bulk of environmental toxins, are spread all over the world, even if the particular chemical is not used in a given area, because of spread via groundwater, surface rain, and winds. Bioaccumulation of these compounds causes disease in living beings; in humans, the immune, endocrine, and neurological systems are the most affected[1].

The effects of toxins are documented in a number of diseases, including asthma, allergies, autoimmunity, cancers, cognitive deficits, and obesity. New diagnoses include the so-called sick building syndrome (SBS) and multiple chemical sensitivity. SBS is a combination of ailments associated with an individual's place of work (typically, but not always, an office building), although there have also been instances of SBS in residential buildings. A 1984 World Health Organization report into the syndrome suggested that up to 30% of new and remodeled buildings worldwide may be linked to symptoms of SBS. Multiple chemical sensitivity is a condition even more difficult to treat and is recognized in some individuals who develop chemical sensitivity to a host of environmental chemicals. Since the WHO report many publications have confirmed these findings[2–7].

## Effects of Toxins in General

Toxins can have several effects on their environment or host, including acute lethality; sublethal effects on plants, nonmammalian species, and mammalian species; teratogenicity; genotoxicity/mutagenicity; and carcinogenicity.

We will consider only the effect on mammals, especially humans. The effect of the toxin on the host will be determined by several factors, namely, the substance and its concentration, the rate of

---

1 Crinnion WJ. Environmental medicine, Part 1: The human burden of environmental toxins and their common health effects. *Altern Med Rev.* 2005;5(1):52–63

2 Burge S, Hedge A, Wilson S, Bass JH, Robertson A. Sick building syndrome: a study of 4373 office workers. *Ann Occup Hyg.* 1987;31(4A):493–504

3 Muzi G, dell'Omo M, Abbritti G, Accattoli P, Fiore MC, Gabrielli AR. Objective assessment of ocular and respiratory alterations in employees in a sick building. *Am J Ind Med.* 1998 Jul;34(1):79–88

4 Stenberg B, Eriksson N, Höög J, Sundell J, Wall S. The Sick Building Syndrome (SBS) in office workers. A case-referent study of personal, psychosocial and building-related risk indicators. *Int J Epidemiol.* 1994 Dec;23(6):1190–1197

5 Nordström K, Norbäck D, Akselsson R. Influence of indoor air quality and personal factors on the sick building syndrome (SBS) in Swedish geriatric hospitals. *Occup Environ Med.* 1995 Mar;52(3):170–176

6 Ooi PL, Goh KT, Phoon MH, Foo SC, Yap HM. Epidemiology of sick building syndrome and its associated risk factors in Singapore. *Occup Environ Med.* 1998 Mar;55(3):188–193

7 Skov P, Valbjørn O, Pedersen BV. Influence of personal characteristics, job-related factors and psychosocial factors on the sick building syndrome. Danish Indoor Climate Study Group. *Scand J Work Environ Health.* 1989 Aug;15(4):286–295

intoxication, and the host. Toxins can cause temporary dysfunction in the body, lead to permanent damage in the case of chronic intoxication, or lead to death in the case of acute intoxication. In reality, all toxins can be detrimental to the body in the relevant dose. In the words of Paracelsus, the dose makes the poison.

The toxicity, or the effect of the toxin on the body in relationship to the dose, is dependent on several factors:

- The galenic form of the toxin. For instance, organic mercury is much more toxic in gaseous form than in liquid form.
- The "potentiation" of a toxin. When a toxin is mixed with other toxins, the combined toxicity may be increased because toxins often potentiate each other. For instance, two environmental toxins may be present in concentrations that are individually subtoxic but together they may become severely toxic.
- The time over which the toxin is taken in (i.e., acute, subacute, or chronic intoxication). For example, if someone drinks 10 liters of water at once, hyponatremia and death will result.
  Often, intoxication in minute amounts over time has a different effect on the organism (e.g., it may cause cancer or affect the immune system), whereas acute intoxication will have a totally different effect.
- Patient variables, including the patient's ability to regulate in view of the intoxication, the sex of the patient, the patient's body weight, and the patient's possible tolerance to the substance, especially if it is given in low doses orally over time. This factor of patient variables has important implications for treatment strategies and will be discussed in greater depth later.

### Measurement of Toxicity

To compare the toxicities of toxins with each other, they are tested under strictly controlled laboratory conditions, mainly on animals. This is expressed as the so-called lethal dose 50 ($LD_{50}$), which determines the dose, per kilogram of body weight, at which half of the test population will die. The dose–response curve of such intoxication is linear or sigmoidal. However, recently, interest has been revived in toxic exposure at doses below the so-called no observed adverse effect level (NOAL) over time.

It is now known that many environmental toxins will exert a stimulatory or an inhibitory effect over time on processes in the body, resulting in the so-called U-shaped response curves, as seen in hormesis. This means that a toxin may only exert a certain effect in a specific low dosage range. If the concentration goes below this very low dose, it has no effect; and if the substance is given at full strength, it may also have no effect or it may have a different effect. Thus, a toxin may be highly toxic when given in a high dose, lose this toxicity at a lower dose, but then still have a detrimental effect over long-term exposure in minute concentrations. This is important, especially for chemical carcinogens acting by a genotoxic effect.[8]

Recently, a publication in *Science* brought to light an even more frightening fact concerning the effect of environmental toxins. This concerns the area of epigenetics, in which genetic material is damaged by toxins in one generation and the damage is transferred to the next generation (i.e., the effects are seen for the first time in the next generation)[9].

The effects on reproduction correlate with altered DNA methylation patterns in the germ line. The ability of an environmental factor (e.g., an endocrine disruptor) to reprogram the germ line and to promote a transgenerational disease state has significant implications for evolutionary biology and disease etiology.

### Classes of Toxins

Given the large number of toxins, it is difficult to classify them chemically, either by function or by mode of action, because many of them will fall into more than one class. Toxins can enter the

[8] Calabrese EJ, La Hormesis B. U-shaped dose responses and their centrality in toxicology. *Trends Pharm Sci.* 2001 June;22(6): 285–291

[9] Anway MD, Cupps AS, Uzumcu M, Skinner MK. Epigenetic transgenerational actions of endocrine disruptors and male fertility. *Science.* Jun 3 2005;308 (5727):1466–1469

body through air, water, and soil pollution (in terms of accumulation in food sources). Others enter the body through deliberate use, such as drugs of abuse, therapeutic drugs, cosmetics, food additives, and contaminants. A number of these toxins can be classified as so-called anthropogenic toxins because they are connected with the actions of humans (e.g., industrial processes, mining and drilling, combustion of fossil fuels for heating and power, and circulation of exhaust fumes from transportation vehicles).

*Atmospheric/aerial Pollutants*

Gaseous pollutants are gases at normal ambient temperatures and pressure. Among the pollutants of greatest concern are carbon monoxide (CO), hydrocarbons, hydrogen sulfide, nitrogen oxides, ozone and other oxidants, sulfur oxides, and carbon dioxide ($CO_2$). Particulate pollutants include dust, fumes, mist, smoke, and aerosols.

Pollutants have numerous effects. One of the most important is the damaging effect on the lung and the respiratory mucosa, which contributes to allergy, asthma, and cancer.

*Metals*

Metals are naturally occurring but accumulate because of their uses in the activities of our societies. They include arsenic, lead, mercury, cadmium, and chromium. Metals have a wide range of detrimental actions on the body and share common toxic mechanisms and sites of action.

- **Enzyme inhibition/activation.** Enzymes, especially those containing sulfhydryl (SH) groups that bind certain metals in the body as cofactors, may be affected because the toxic metal could displace the cofactor.
- **Subcellular organelles.** Toxic metals disrupt the structure and function of a number of organelles, such as the endoplasmic reticulum, the lysosomes, and the mitochondria. Metal inclusion bodies may form in the nucleus.
- **Carcinogenicity.** Arsenic, chromium, nickel, cadmium, and cisplatin are human carcinogens. This carcinogenicity arises because of the interactions of the metallic ions with DNA.
- **Kidney.** The main excretory organ in the body for metals is the kidney. Cadmium and mercury are especially nephrotoxic.
- **Nervous system.** The lipid-soluble metal compounds, such as methylmercury and organic lead compounds, are neurotoxic because they readily cross the blood–brain barrier.
- **Endocrine and reproductive effects.** Because the male and female reproductive systems are under complex neuroendocrine control, any toxin interfering with any of these processes can affect the reproductive system.
- **Respiratory system.** Short-term exposure to metals leads to irritation and inflammation of the respiratory tract, whereas long-term exposure may result in fibrosis (aluminum) or carcinogenesis (arsenic, lead, and nickel).
- **Metal-binding proteins.** The toxicity of many metals, such as cadmium, lead, and mercury, depends on their transport and intracellular bioavailability. This is regulated by the high affinity for certain cytosolic proteins. These proteins are typically rich in thiol (SH) groups. These intracellular "sinks" are particularly important for sequestering metals away from vital organelles. The ECM must also be mentioned as it will be an important storage place for metals before they reach the cell because it is also rich in thiol groups.

*Agricultural Chemicals (Pesticides) Including Wood Preservatives*

The use of chemicals to control pests goes back centuries to the ancient Chinese and the Romans. However, in the 20th century, the types of compounds that we now identify as pesticides were developed. Ideally, pesticides should be highly specific and only harm the intended target; however, many health hazards have been recognized as originating from these substances. A major risk is environmental contamination, especially where they may enter the food chain and natural water resources. A very serious concern is the long half-lives of certain of these compounds, as well as their lipophilic nature, which makes bioaccumulation and especially accumulation in the human body a real threat. Many of these compounds are classified as persistent organic pesticides.

The globally harmonized system for the classification and labeling of hazardous chemicals is an initiative to promote common, consistent criteria for classifying chemicals according to their health, physical, and environmental hazards and to develop compatible labeling, safety data sheets for workers, and other information based on the resulting classifications. This is an ongoing initiative of the Environmental Protection Agency in the United States.

The primary classes of pesticides produced are fumigants, fungicides, herbicides, and insecticides. In the United States alone, the figure for the total annual production of such chemicals is 1.2 billion pounds (544310.844 metric tonnes). Some 665 million pounds (301638 metric tonnes) of wood preservatives are also produced.

Organochloride pesticides are probably the best-known class of pesticides. These substances, introduced in the 1940s and 1950s, include familiar pesticides such as dichlorodiphenyltrichloroethane (DDT), chlordane, and dieldrin (developed later as an alternative to DDT).

In acute intoxication, organochlorides act as neurotoxins. The persistence of pesticides of this group is a major concern, and they were banned in the United States in 1972, 10 years after the publication of Rachel Carson's *Silent Spring.*[10] However, groundwater in many parts of the world remains contaminated.

Biological effects of pesticides and wood preservatives include direct damage to DNA (thus, carcinogenic effects), interference with immune function, induction of the P450 enzyme detoxification system, and endocrine disruption.

### *Plasticizers and Polybrominated Biphenyl Ethers*

Plasticizers, most commonly phthalates, alkyl phenols, and bisphenol A, are used to soften the materials to which they are added (e.g., those that give a hard plastic like polyvinyl chloride its flexibility). The polybrominated biphenyl ethers (PBBs) of particular concern include the so-called phthalates occurring in softeners of plastics, oily substances in perfumes, additives to hairsprays, lubricants, and paint and wood finishers. Flame retardants such as PBBs are materials that inhibit or resist the spread of fire.

Although high doses of phthalates do constitute risks in the sense of traditional toxicology, low doses also can have dramatic effects. Research reveals that male reproductive development is intensely sensitive to some phthalates. For example, two phthalates, dibutyl phthalate and diethylhexyl phthalate, produced dramatic changes in male sexual characteristics when exposure took place in utero, at levels far beneath those of previous toxicological concern. These changes included increases in the rates of hypospadias and indications of demasculinization[11].

Bisphenol also has been implicated in the development of cancer.[12]

### *Therapeutic Drugs and Drugs of Abuse*

Therapeutic drugs and drugs of abuse constitute a huge group of chemicals that will not be discussed in depth here. Apart from the known side-effects of this group, many will also disrupt the self-regulatory processes of the body and will cause deregulation and other disease. Many who will treat one condition will put the patient at risk of another. For instance, many immunosuppressant agents, used in autoimmune diseases, are known to be carcinogenic.

### *Food Additives and Preservatives*

Compounds are added to foods as preservatives (such as methyl-*p*-benzoic acid and propionates) and other additives are used to change the food's physical characteristics such as color (tartrazine), odor, and taste (saccharin and aspartame, pipero-

[10] Carson R. *Silent Spring.* Greenwich, CT: Fawcett Crest; 1962

[11] Clark, R. In Daston, G, & Kimmel, C (Eds.). (1999). Endpoints of reproductive system development. *An Evaluation and Interpretation of Reproductive Endpoints for Human Health Risk Assessment.* Washington, DC: ILSI Press

[12] Welshons WV et al. Large Effects from small exposures. I. Mechanisms for endocrine-disrupting chemicals with estrogenic activity. *Environ Health Perspect.* 2003;111(8):994–1006

nal, and nitrates and nitrites). Hundreds, and possibly thousands, of food additives are used worldwide, many with inadequate testing. Some of these were introduced when toxicity testing was not sophisticated and were subsequently were shown to be toxic. The question of synergistic interactions between all of these substances has not been looked at adequately. Many of these substances are carcinogenic or mutagenic.

*Endogenous Toxins*

Endogenous toxins are products of normal regulatory processes in the body that accumulate, either via overproduction or because they are not adequately metabolized or excreted. Substances such as adrenaline and histamine are examples.

These substances are also metabolized by the same processes that metabolize external toxins. Overloading of these enzyme systems by external toxins can lead to the accumulation of these internal toxins. This is why, for instance, patients with toxic reactions will often have panic attacks, because the metabolism of adrenaline will be impaired. Psychotoxins will cause an imbalance in the neurotransmitter status and will disturb normal homeostasis. Furthermore, cortisol, the chronic stress hormone, plays a very important role in the metabolism and the resultant detoxification of the ECM, so that chronic stress will impair the renewal of the matrix and promote an accumulation of toxins in the matrix.

*Conclusion*

- Toxins are ubiquitous in our environment and can accumulate for years in the groundwater, as well as cover vast distances because of atmospheric transport.
- There are different classes of toxins, such as atmospheric/aerial pollutants, metals, agricultural chemicals, plasticizers, food additives, therapeutic drugs, and endogenous toxins (which are formed in the body and not adequately excreted).
- Many toxins occur at the so-called NOAL but still cause long-term problems, such as carcinogenesis, if present at minute concentrations.
- Epigenetic effects of toxins are now recognized, in which one generation is exposed to the toxin but the effect is only seen in the offspring.
- The toxicity of toxins ranges from lethality to teratogenicity to genotoxicity and carcinogenicity.
- Toxins can also impair the immune system and the intracellular organelles, can cause enzyme inhibition or activation, and can mimic endogenous hormones, when they are called hormone disruptors.

A practical classification of toxins can be found in **Fig. 4.1**.

### The Physiology of Detoxification

In the previous section, we saw that the body is exposed to a wide variety of toxins present in the air, water, soil, and food. The body has several inherent defense mechanisms and membrane barriers to prevent the absorption and distribution of the toxin when the intoxication has occurred. Once the toxins are inside the body, the internal defense system, or basic bio regulating system (BBRS; see Chapter 1), will be mobilized to eliminate the toxin or to at least try to compensate for it.

Here we will look at the way the body deals with these toxins. Before a toxin can have a detrimental effect on the body, it needs to reach the target organ or cell. In principle, a toxin can go through the following five stages:

1. Absorption
2. Transport
3. Distribution and storage
4. Metabolism
5. Elimination

*Toxicokinetics* studies the absorption, distribution, elimination, metabolism, and/or clearance that take place in the body after exposure to the toxin. *Toxicodynamics*, on the other hand, studies the biochemical and physiological effects of drugs and toxicants and determines their mechanism of action. Toxicokinetic processes are schematized in **Fig. 4.2**.

## EXOGENOUS ("Toxicants" or "Xenobiotics")

### INGESTION

The mucosal surface of the GI tract is about 200 x that of the skin surface, in a person's lifetime over 25 tons of food is processed by the GI system, thus an enormous load of possible toxins (antigens, xenobiotics, microbes etc.)

#### FOOD

**Chemical Contamination**

1 **Toxic Metals**
Arsenic, Lead, Cadmium, Hg, etc.
2 **Polycyclic Aromatic Hydrocarbons**
from incomplete combustion of hydrocarbons
3 **Industrial Chemicals**
PCBs
Chloroform
Trichloroethylene
etc.
4 **Hormones and Drugs in Animal**
5 **Fertilizers**
6 **Pesticides**

**Microbes**

1 **Bacteria**
viruses
2 **Protozoa**
e.g. Giardia

**Mycotoxins**

1 **Toxins Produced by Molds**
e.g. Aflatoxins produced by the *Aspergillus* molds

**Food Additives**

1 **Colorants**
2 **Preservatives**
etc.

#### WATER

**Chemicals**

1 **Solvents**
2 **Phosphates**
3 **Nitrates**
4 **Herbicides**
5 **Pesticides**
6 **Fertilizers**
7 **Industrial wastes**
etc. ... Water is usually analyzed for fewer than 60 of over 700 chemicals found regularly in drinking water.

**By-Products of Microbes**

1 **Bacterial**
e.g. *E. coli*
2 **Viruses**
e.g. Hepatitis virus
3 **Parasites**
e.g. *Giardia*
4 **Algae**
and their toxins

**Heavy Metals**

1 **Mercury**
2 **Lead**
3 **Arsenic**
etc.

**Others**

1 **Asbestos**
2 **Radioactive elements**
radon, radium, uranium
3 **Gasoline**
etc.

#### DRUGS

Prescription

Recreational

### INHALATION (Environmental)

The lungs have the greatest exposure of any organ to the environment. The air we breathe contains dust, chemicals, pollutants, gases, microbes, small particles and liquid aerosols.

#### OUTDOOR

Air quality standards measure 6 pollutants:

1 **Suspended particulates**
2 **Carbon dioxide**
3 **Nitrogen oxides**
4 **Sulfur dioxide**
5 **Photochemical oxidants**
e.g, ozone, aldehydes
6 **Lead**

**Natural Sources of Air Pollutants**

1 **Volcanoes** (ashes)
2 **Natural gas**
3 **Terpenes** (plants)
4 **Ammonia**
(from biological decomposition)
5 **Smoke** (fires)
6 **Dust** (soil)
7 **Plants/pollens**
8 **Microbes**

**Human-Caused Air Pollutants**

1 **Chemical dumps**
2 **Waste disposal**
3 **Fuel combustion**
4 **Transportation**
5 **Industrial**
6 **Farm spraying**
etc.

#### INDOOR

Indoor air pollutants may come from outdoor, from materials in the building, or from human activities.

**Chemicals and Minerals**

1 **Asbestos**
2 **Formaldehyde**
3 **Volatile organic compounds (VOCs)**
4 **Radon gas**
5 **Nitrogen oxide**
6 **Carbon dioxide**

**Furniture and Renovations**

1 **Wood** (phenols & formaldehyde from plywood, paelling, etc.)
2 **VOCs** (from glues, fillers, paints, varnishes, etc.)
3 **Paints** (with volatile fungicides, pesticides, mildew-cides)
4 **Fiberglass** (from insulations)
5 **Plasticizers** (flexible vinyl floors)
6 **Upholstery** fabrics & carpets
(dye, formaldehyde, plasticizers, fungicides)
7 **New carpets** (contain more than 20 chemicals to kill bacteria, hold colors, bind fibers and also release acetone, benzene, styrene, xylene, toluene and formaldehyde ... in addition to dust, chemcials, and microbes that they can harbor)

**Household Products**

1 **Personal care products**
2 **Laundry products and fabric softeners contain numerous toxic chemicals such as:**
**Carcinogenics** (chloroform, benzyl acetate, limonene)
**S.N.C. toxins** (camphor, ethyl acetate, benzyl alcohol, linalool, pentane)
3 **Household cleaning products**
4 **Pesticides** (used frequently)

**Microbes, Molds, Dust, Pets**

1 **Molds and mildews in human areas**
2 **Dust and dust mites**
3 **Bacteria, viruses, fungi, etc.**
4 **Pets increase toxins** (dander, fleas, use of flea powder and collars that have toxic chemicals, etc.)

**Human activities**

1 **Transmission of microbes**
2 **Tobacco smoke and fireplaces**
3 **Recreational drugs** etc.
Air-conditioning and heating systems together with "better" sealings from windows/doors have drastically reduced natural ventilations—the number of air exchanges has been practically reduced to zero (toxins remain and further accumulate inside)

### DERMAL (Skin)

**Active (Injections)**

1 **Prescription drugs**
2 **Recreational drugs**
3 **Animal toxins**
bites or puncture by fish, arhropods, parasites, etc.

**Passive**

Substances that are both water and fat soluble are more easily absorbed through the epidermis especially if not intact and through the hair follicles:
1 **Drugs**
2 **Cosmetics**
3 **Chemicals**
especially from the air and from waters ... showers, bathing, etc.)
4 **Radiations**

## ENDOGENOUS ("Toxins")

**Produced in the Body**

1 **Physiologically**
– **Bilirubin**
– **Ammonia**
– **Uric acid**
– **Lactic acid**
– **Creatinine**
etc.

Become "toxic" if in excess for:
– production
– detoxification and excretion

2 **Under Abnormal Conditions**
– **production of waste products** ($CO_2$, $H_2O_2$, free radicals, etc.)
– **hormones and/or neurotransmitters**
– **microbial debris**
– **pH imbalances**
etc.

**Stored in the Body**

Originally from external origin but introduced into the body where they are stored and become a continuous source of "toxic" release (Water-soluble chemicals are easily excreted, but fat-soluble chemicals accumulate in fat cells and cell membranes)

1 **Dental materials**
2 **Medical implants**
3 **Microbes (foci)**
etc.

**Fig. 4.1** Practical classification of toxins. (There is no place in which we do not encounter toxins, but by becoming aware of the exposure possibilities, we may better plan to avoid them, and if that is not possible, at least to limit them and/or learn to detoxify them.)

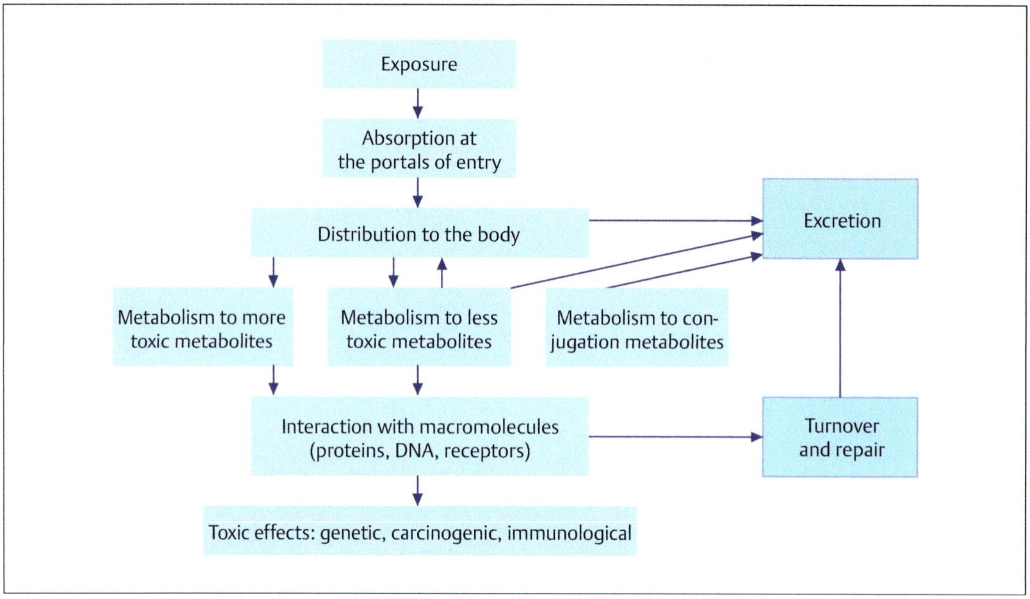


**Fig. 4.2** Toxicokinetic processes.

### *Absorption*

Toxins can enter the body through all the surfaces that are in contact with the outside world. These surfaces comprise the skin, the mucous membranes, and the gastrointestinal tract (GIT). In general, absorption via the respiratory mucosa is the quickest, and absorption via the dermal route is the slowest. The extent of absorption depends on the amount of toxins present externally and on the saturability of the transport process.

The mucosal surfaces have several barriers that will prevent toxins from entering the body, such as a mucosal barrier, the physical presence of symbiotic bacteria, and the so-called tight junction. The skin also has a barrier in the form of a certain level of pH.

### *Transport*

Once the toxicant is absorbed into the body, it moves around in two ways: by bulk transfer via the blood or lymph, and locally by diffusional transfer over short distances. The path that a toxin takes after absorption is illustrated in **Fig. 4.3**.

During absorption, distribution, and elimination, a toxin will encounter various cell membranes before interacting with the target tissue. These membrane barriers will differ from relatively thick areas of the skin to relatively thin lung membranes, although their composition is essentially similar.

The cell membrane can be seen as a lipid matrix. It contains both phospholipid (hydrophobic) portions and hydrophilic heads. Transmembrane proteins are embedded in the membrane. The lipid layer will differ from organ to organ. The myelin in the brain consists of 100% lipid bilayer, whereas mitochondria have only a 40% lipid bilayer. This, of course, has implication for the distribution of fat-soluble toxins. The fat solubility of the toxin determines how well it will traverse the cell membrane. Many of the proteins that traverse the membrane are active in the transport of toxins over the cell membranes.

The distribution of the absorbed toxin will depend on physiological factors and on the physiochemical properties of the toxic substance. The process is, thus, a reversible movement of the toxicant between the blood and tissues and between the extracellular and intracellular compartments.

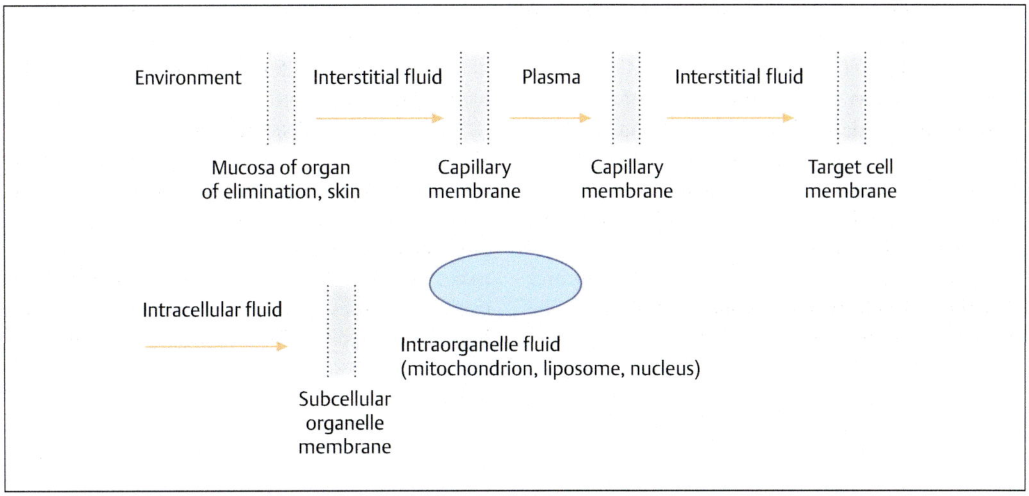


**Fig. 4.3** Toxin pathways after absorption.

The rates of these reversible movement become important when we address the mobilization and drainage of toxins later.

Perfusion and protein binding are two of the factors that can complicate the distribution of a toxin:

**Perfusion of the organ.** The well-perfused tissues include the liver, kidneys, and brain; the poorly-perfused tissues include the bone and fat tissue, from which there is only slow elimination of toxins.

**Protein binding.** There may also be significant protein binding that can affect the delivery of the drug to the tissues and vice versa. In particular, binding of toxins to the matrix structures may trap these toxins for years and prevent their elimination.

Protein binding plays a very important role in the transport of toxins. There are many plasma proteins involved in such transport: albumin, α-acid-glycoprotein, the lipoproteins, and globulins are mostly involved.

The lipoproteins, such as high-density, low-density, and very low-density lipoproteins, are important because so many toxins are lipophilic; the lipoproteins will therefore carry a number of toxins. Iron and copper will be carried by the metal-binding globulins transferrin and ceruloplasmin.

*Distribution and Storage*

Plasma protein binding is not selective and toxins can compete with each other and even with endogenous substances for binding. Covalent binding to the protein forms a minor part, but the dissociation of covalent bonds is extremely difficult and the carrier molecule is changed and may eventually play a role in carcinogenesis. Noncovalent binding is more common. The toxin can dissociate more easily from this bond. However, in some cases, the bond may be so strong that the toxin remains bound for weeks, months, or years. Certain metals have high association constants and their dissociation is extremely slow.

If the affinity for an organ is high, the toxin will accumulate or form a depot for years. In general, lipid-insoluble toxicants stay in the plasma and interstitial fluids, whereas lipid-soluble contaminants reach all compartments and may accumulate in fat.

Some toxins have a specific affinity for certain tissues. Tetracyclines have a high affinity for the calcium-containing tissues, which is seen in the discoloration of teeth if they are given when an individual is younger than 14 years. Similarly, chloroquine, an antimalarial agent, has an affinity for melanin and can be taken up by tissues such the retina, causing retinitis. This drug is often used in patients with lupus and other connective-tissue diseases, which makes an ophthalmic checkup every 6 months mandatory.

Bone will also contain concentrations of certain toxicants, such as lead. A sudden loss of bone can lead to the sudden release of the toxin and can have dire consequences, especially after menopause, when there may be sudden bone loss. This will be discussed later when we look at the ideal rates of detoxification.

Lipophilic pesticides, such as the organochlorines and the polychlorinated biphenyls (PCBs), can be expected to accumulate in fatty tissues. The affinity of metals for SH groups was addressed in earlier sections. The binding of these metals to the numerous thiol groups in the ECM is of special concern.

Certain areas will be naturally less penetrable by toxins. The brain, which is protected by the blood–brain barrier, is an example. Disease processes, such as meningitis and other inflammatory or infective conditions, can disrupt this barrier and can cause toxins to enter the brain tissue. Other tissue–blood barriers include the prostate–blood barrier and the testis–blood barrier.

Unfortunately, the placenta is a poor barrier and the fetus is exposed to all the toxins to which the mother is exposed. This was seen in fatty-tissue biopsy specimens that were obtained from newborns, in which numerous toxins, such as PCBs and dioxin, were found. Therefore, because of today's environmental pollution, we need to assume that newborns are already contaminated with toxins.

### *Metabolism*

One of the most important determinants of the persistence of toxins in the body is the extent to which they can be metabolized and excreted. Several families of metabolic enzymes are active in the metabolism of endogenous and exogenous toxins. These include the P450 system, which is one of the most important, as well as the flavin-containing monooxidases (FMOs), the alcohol and aldehyde dehydrogenases, amine oxidase, COXs, reductases, hydrolases, and the conjugating enzymes, such as the methyltransferases and the glutathione transferases.

Most of the metabolism takes place in the liver. Because most of the toxins entering the body are lipophilic, they need to become water-soluble for excretion. After entry to the liver and other organs, xenobiotics may undergo two phases of metabolism, demonstrated in **Fig. 4.4**.

Phase I metabolism involves mainly the cytochrome P450 (CYP) system, the FMOs, and the hydrolases. After the addition of a polar group, conjugating enzymes typically add more constituents, such as sugars, sulfates, or amino acids that make the compound more water-soluble. However, sometimes more toxic intermediate metabolites are formed via this process. An example is the breakdown of alcohol to acetaldehyde, which is much more toxic than alcohol.

The CYP system plays a very important role in phase I reactions. The CYPs, which constitute the CO-binding pigment of the liver microsomes, are heme proteins. A nomenclature system has been developed for the different types and isoforms. Although mammals are known to have 18 CYP families, only three are responsible for xenobiotic metabolism. The remaining families are involved in steroid hormone production. They are designated according to the gene, subfamily, and isoform (Arabic numeral, letter, and Arabic numeral).

Thus, CYP 3A4 is responsible for the metabolism of many drugs and endogenous and exogenous toxins. Its activity can also be influenced by a host of drugs and chemicals. It can be induced, meaning that certain drugs are broken down too quickly (e.g., warfarin). In contrast, consumption of grapefruit juice in large quantities is known to damage this system irrevocably and may lead to an accumulation of drugs.

The phase I detoxifying pathway takes care of environmental toxins (e.g., pesticides, pollutants, and food additives) and drugs and alcohol. The end products of our own metabolism are also processed here for excretion. Fat-soluble toxins are changed by oxidation, reduction, and hydrolysis to make them more water-soluble for excretion via the bile and the kidney. The phase I enzymes need certain cofactors to fulfill their action. These are trace elements, vitamins, amino acids, and substances such as the reduced form of nicotinamide–adenine dinucleotide (NADH).

Phase I produces significant amounts of free radicals during this detoxification process. If the antioxidant status of the patient is not adequate, tissue damage may occur if the P450 system is

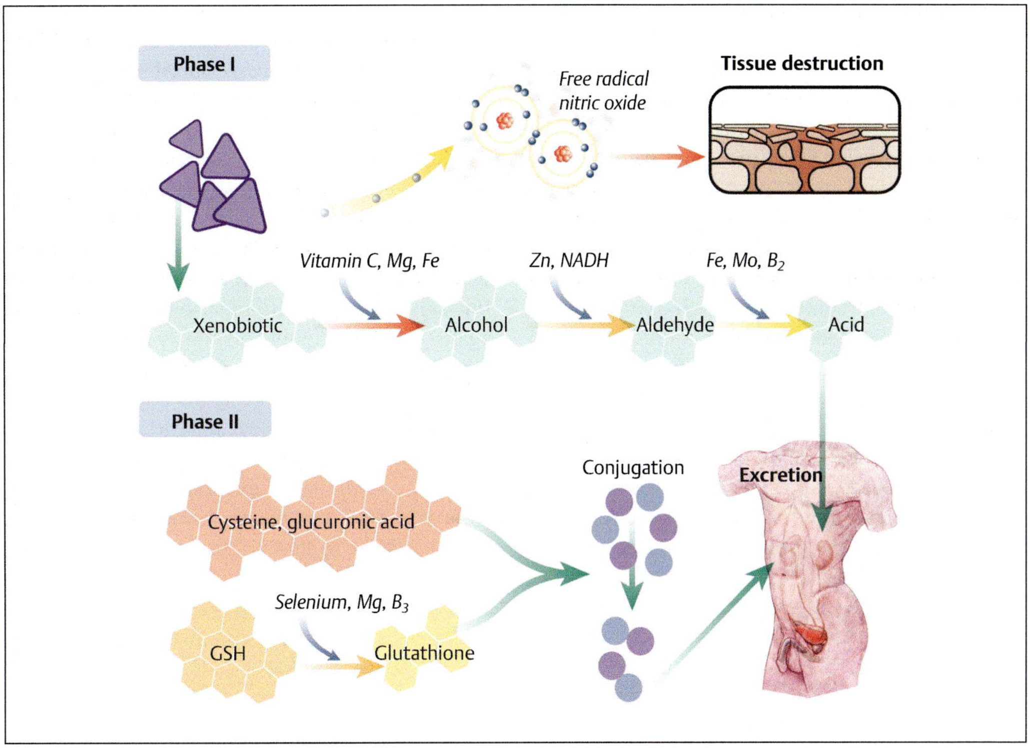


**Fig. 4.4** Liver metabolism pathways.

overloaded or induced. Some substances, such as caffeine, alcohol, certain drugs, dioxin and organophosphates (used as pesticides), and paint fumes, can induce this pathway. Sometimes, intermediate substances such the acetaldehyde formed during the metabolism of certain toxins, such as alcohol, can be more toxic to the body than the original substance. Certain people, called fast acetylators, will be more prone to damage of the liver because the toxin is metabolized quickly to this dangerous intermediate and then the process becomes slow again. These individuals are at higher risk for liver damage during ingestion of toxins. The alcohol dehydrogenase pathway is used to detoxify these substances (e.g., in the case of acetaminophen overdose).

The phase II pathway, or the conjugation pathway, uses substances rich in SH groups to metabolize toxins. A number of these substances, such as cysteine, taurine, and glutathione, which are formed from glycine, glutamine, and cysteine via a selenium-dependent enzyme, act as free-radical scavengers and heavy-metal chelators. During conjugation of toxins, they are lost to the body forever because they are excreted with the toxin; as free radicals, they can be regenerated.

Some substances will require only either phase I or phase II to be detoxified, whereas others will require both phases. Thus, it is clear that if the phase II pathway is overloaded, the free-radical scavenging ability will be given up in favor of the conjugation function and further damage to the liver parenchyma may occur. For example, if the patient is deficient in selenium, glutathione production will be impaired, resulting in toxicity and free-radical damage.

### *Elimination*

After the toxins have gone through these two phases, they are ready to be eliminated. However, if the intermediary toxin is not broken down, or if the toxin load is too high, there will be bioaccumulation of the toxin.

The ability to detoxify and eliminate toxins is paramount to the maintenance of health in an organism. For unicellular organisms, a simple process of diffusion is enough to eliminate toxins; however, multicellular organisms, especially those with more complex structure, need to find other ways to eliminate toxins. With an increase in complexity of organisms comes an increase in size, a decrease in surface area to body mass ratio, compartmentalization of cells and organs, and an increase in lipid content. Together with the fact that organisms need to protect themselves from the environment with barriers such as scales and skin, there is less possibility for toxins to diffuse out of the body. This problem was solved by developing specialized methods of metabolism for toxins and specialized routes of elimination. Thus, we have major and minor elimination routes. The major routes involve the liver, kidneys, mucous membranes, lungs, and skin, whereas the minor routes involve the saliva, sweat, milk, hair, and secretion from reproductive organs.

For elimination of the toxin, it must go through the reverse of the route described for absorption, from the place of storage back to the external environment (**Fig. 4.5**). Compounds are transported from the place of storage mainly via the bloodstream. The circulatory system lends itself to the transport of water-soluble substances. The more lipophilic a substance is, the less likely it is to freely diffuse into the blood; thus, mobilization of these toxins from their place of storage is more difficult. The same process discussed for absorption—binding of toxins to carrier proteins and lipoproteins—is the way these toxins will enter the bloodstream.

The toxins are thus transported back to the organs of elimination. However, if these organs are dysfunctional, overloaded, or damaged, the toxins cannot be excreted. This means that such toxins will circulate further in the bloodstream and through diffusion will enter some compartments again (e.g., a fat-soluble toxin may be stored in the brain, with dire consequences). Therefore, stimulation of the movement of toxins out of their compartments has to be a slow and careful process.

Two groups of compartments should be distinguished: (1) the *rapid-exchange system*—in these compartments, the tissue concentration of toxicant is similar to that in blood; and (2) the *slow-exchange system*—in these compartments, the tissue concentration of toxicant is higher than in the blood because of binding and accumulation of adipose tissue (the skeleton and kidneys can temporarily retain some toxins, such as arsenic and zinc).

Detoxification and drainage should continue until the slow-exchange system is given a chance to excrete all the toxins. The organs involved in the detoxification and drainage of the toxins will be discussed in the next section.

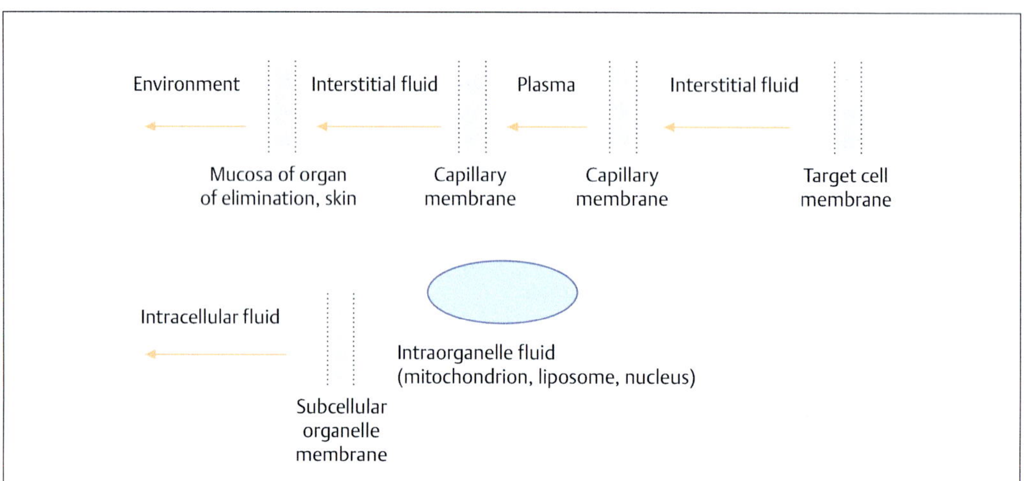


**Fig. 4.5** Toxin pathways for excretion.

*Conclusion*

- Toxins have to cross several membranes in the body in order to be absorbed and eventually to be stored or eliminated via the organs of elimination.
- Many toxins follow simple kinetics (diffusion over semipermeable membranes occurs until a steady state is achieved on both sides of the membrane).
- These basic kinetics are different for toxins with a high association coefficient with proteins and cellular structures, be it in the blood or in the organ of storage.
- These kinetics affect both the storage and the mobilization of toxins in and out of these compartments and must form the basis on which practical detoxification and drainage is executed.
- Two groups of compartments can be distinguished, depending on the perfusion of the organ and the amount of toxin bound to protein: (1) compartments in which the tissue concentration of toxin is similar to that of the blood (rapid-exchange system) and (2) compartments in which the tissue concentration of toxin is higher than in blood because of binding and accumulation (slow-exchange system).
- Many toxins are metabolized before they can become excreted. One of the main purposes of this is to render fat-soluble toxins water-soluble for excretion in the bile and kidneys.
- The P450 system of enzymes plays a major role here, especially in the liver, where it comprises the phase I reactions; this is augmented by the phase II reactions.
- Organs are in danger during detoxification and drainage because of the high concentration of toxins moving through the organ at the time and the generation of free radicals during the detoxification process.
- Support of these organs is most important during detoxification and elimination of the toxin.

## The Organs of Detoxification and Elimination

As described in the previous section, many organs and tissues are involved in the absorption, transport, metabolism, storage, and elimination of toxins. These highly complex processes require special properties of the organ to fulfill this function. Often, the organ itself may be endangered by disease because of the excretion, storage, and movement of toxins through it. These organs will be described in more depth.

### *The Liver*

The liver is one of the body's most important detoxification and elimination organs. Metabolically, it is the most complex. The liver is a major organ of chemical elimination in that it takes up compounds from the blood, metabolizes them, and ensures the biliary and renal secretion of toxins. The liver detoxifies a large array of external and internal toxins. It also plays a role in cholesterol metabolism, glycolysis, and gluconeogenesis, providing many of the plasma proteins necessary for carrying hormones and fats. Finally, the liver provides clotting factors.

Most toxins reach the organs of elimination via the bloodstream. The liver is very well perfused and gets its blood from two sources: (1) the arterial oxygen-rich blood that is delivered through the hepatic artery, and (2) the venous blood through the portal vein, from which the liver gets all the blood that is shunted from the capillaries of the gut and spleen.

The principal cell in the liver responsible for the detoxifying action is the *hepatocyte*, which facilitates the two pathways discussed above (see **Fig. 4.4**) in dealing with mainly fat-soluble toxins to render them hydrophilic or water soluble. Hepatocytes are bathed in blood as this blood transverses a system of sinusoids. This provides a very large surface across which chemicals can easily diffuse into the hepatocytes (see **Fig. 4.6**). Because of the highly lipophilic character of many of the chemicals that are metabolized by the liver, they require carrier proteins to be able to enter the water-soluble area. Several intracellular carrier proteins are present in hepatocytes.

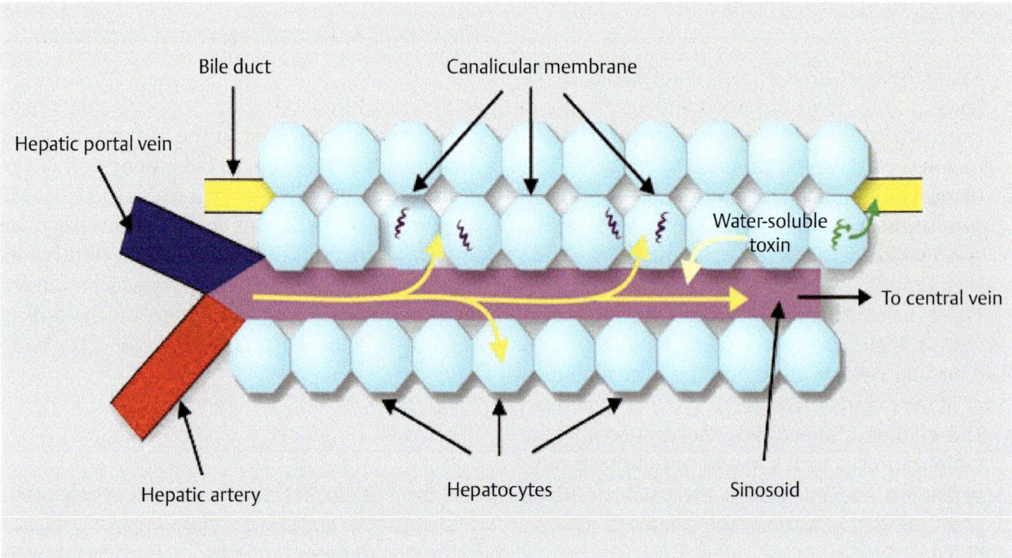


**Fig. 4.6** The liver as an organ of detoxification.

Once inside the hepatocyte, the chemical can interface with the phase I and II enzymes to undergo biotransformation and become water soluble. A number of these substances then diffuse back into the blood, where they will be transported to the kidneys for elimination.

### Excretion in Bile

The biotransformed molecules can also diffuse across the membranes of the bile canaliculus and, thereby, flow into the bile duct. The substance is then further delivered with the other constituents to the gallbladder, which excretes the bile into the intestine for fecal elimination (**Fig. 4.6**).

In many instances, we want to facilitate the drainage of bile. The gallbladder's primary function is to secrete bile and release it through the cystic duct. This duct joins the hepatic duct from the liver to create the common bile duct, which then empties into the upper part of the small intestine (i.e., the duodenum). Bile carries away and neutralizes toxins; it also stimulates and aids digestion by emulsifying fats, stimulating peristalsis, and acting as a natural laxative.

### Enterohepatic Circulation

In this process, already conjugated chemicals, which are water soluble, are deconjugated by hydrolytic enzymes in the gut and rendered lipophilic again. They are then once more reabsorbed by the gut. Thus, the liver is exposed to another round of the same toxin for reprocessing again and again. This increases the retention time for toxic chemicals in the liver and may increase liver toxicity. Some of these metabolites are more dangerous than their original substance. As previously noted, the P450 system is also a source of free radicals that will further damage the liver cell.

Thus, it is imperative to protect the liver during the process of detoxification and drainage.

### The Kidneys

The kidneys specialize in the excretion of numerous water-soluble toxins and metabolites, maintaining homeostasis of the organism. The kidneys detoxify drugs, heavy metals, and other toxins.

Each kidney possesses about 1 million nephrons that perform excretion. Renal excretion represents a very complex event encompassing three different mechanisms: (1) glomerular filtration by the Bowman capsule, (2) active transport in the proximal tubule, and (3) passive transport in the distal tubule.

Blood is delivered to the kidneys via the renal artery. About 625 mL of plasma moves through the kidneys per minute; of that amount, 125 mL is filtered through the glomerular membrane. Most of the water is then reabsorbed in the proximal and distal tubules, so that only approximately 1 to 2 liters of urine are formed per day (see **Fig. 4.7**).

Some filtered substances, such as glucose, will be totally reabsorbed, so that in normal conditions there is no glucose in the urine. Other substances, many of which are harmful to the body, are filtered, secreted, and then minimally reabsorbed. Creatinine is such a substance and can, thus, be used to test the efficiency of the kidneys in clearing harmful substances. It accumulates in the blood when the kidneys are dysfunctional. If the kidneys are damaged through disease or intoxication (drugs and chemicals), their ability to excrete drugs is reduced; in conventional medicine, the dose of drug needs to be adjusted accordingly.

It is important for urine to be on the alkaline side because it facilitates the secretion of certain drugs, such as barbiturates, and alkaline urine will prevent urinary tract infections.

The kidneys commonly bear the brunt of chemical toxicity because the nephron tends to concentrate the toxin and, thus, increase levels of toxic exposure in the tubules. Thus, the kidneys also need protection and support throughout the detoxification and drainage stages.

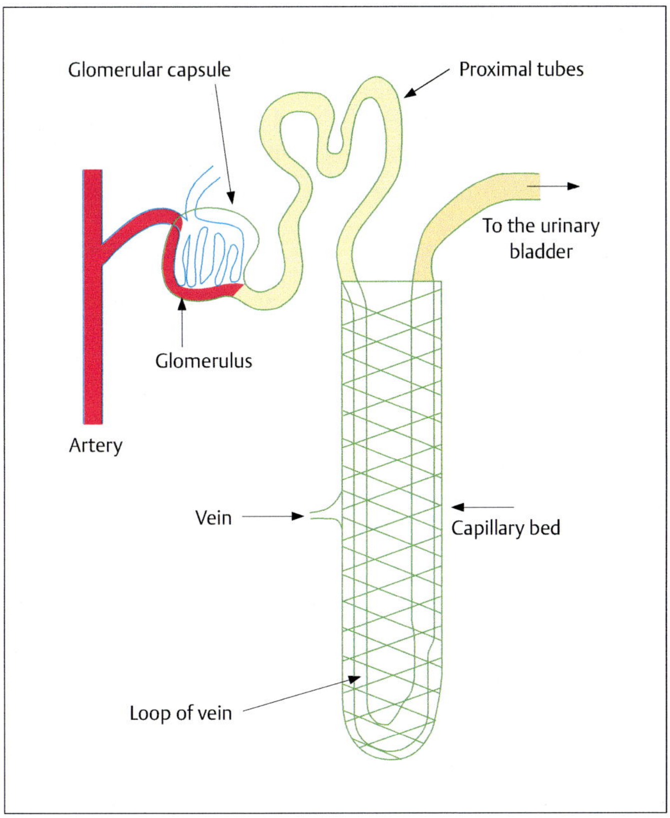


**Fig. 4.7** The kidney as an organ of detoxification and elimination. Structure of the nephron.

*The Matrix and Lymph*

### The Matrix

The matrix forms the final biophysical layer between the cell and the regulatory organs. This system has largely been forgotten since Virchow, a physician who worked in Vienna and was a contemporary of Sigmund Freud, observed a cell through a microscope and postulated that all diseases originated on the cellular level. Another physician working there at the time, Rokitansky, attempted to introduce the humoral theory but was largely ignored. Pischinger and Heine, two modern researchers, extended these ideas; the newer molecular biology texts increasingly recognize the role of the matrix.

The cell on its own is actually an abstraction. The cell does not come into contact with the blood vessels, nerves, veins, and lymph vessels that deliver nutrients and messengers and remove toxins. It relies on the biophysical layer, which is composed of highly polymerized sugar–protein complexes called glycosaminoglycans (GAGs), such as hyaluronic acid, chondroitin sulfate, and heparin. When they are linked to a protein backbone, they are called proteoglycans (PGs). This molecular sieve must be crossed by the entire array of metabolic products. Unfortunately, because of the chemical and electrical charges on the GAGs and PGs, they also become the place where toxins are stored for a long time. The matrix is one of the tissues with slow perfusion and, thus, will have a pattern of slow turnover.

Sugar–protein complexes are phylogenetically considered the best carriers of information. Heine and Pischinger showed that if the matrix is disturbed by a pinprick in one place, the disturbance is communicated to the whole matrix within seconds. This makes it an ideal system through which to impart any information to the body. The acupuncture point is an anatomical structure originating in the matrix, a bell-shaped structure, and it offers a wonderful "window" into this system.

The matrix (see **Fig. 4.8**) has its own biorhythm that is dependent on cortisol and thyroid hormone for activation. During the early hours of the morning, the body goes into an ebb phase with a low cortisol level; it is during this ebb phase that the matrix will purge itself of toxic materials. Stressed patients, or patients who through a change in their sleep–wake cycle have disturbed the diurnal rhythm of cortisol, will not be able to undergo detoxification because there may be a "misfiring" between the matrix and the liver. Cortisone in high doses as a medication will also disturb the innate rhythm of the body, resulting in matrix tox-

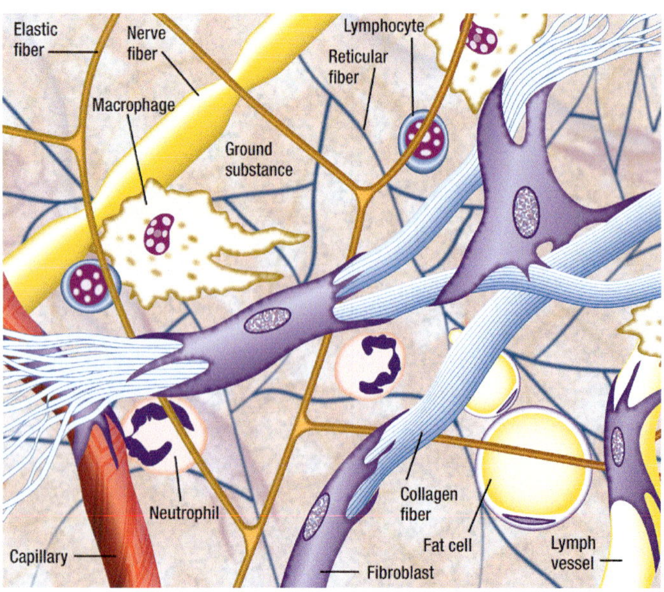


**Fig. 4.8** The extracellular matrix (ECM).

icity. In patients who have been receiving cortisone therapy, swelling and puffiness can be seen in the matrix. Patients who are hypothyroidal have been described as having "myxedema" in older literature. The same swelling will be apparent in the matrix if the matrix biorhythm is disturbed. Many toxins are hydrophilic and will draw fluid into the matrix. The result is edema, which is seen as cyclical edema in females or as cellulite in clinical medicine.

It is clear that if the molecular sieve of the biophysical layer fails (i.e., if it is polluted), there will be distortion of information to and from the cell. If the disturbance is severe enough, cellular disease will ensue.

Newer molecular biological research shows that the matrix is the site for many messengers that code for intracellular phenomena, which, if disturbed, can contribute to many diseases, including cancer[13].

### The Lymph System

In addition to its role in the immune system, the lymph system acts as a detoxifying organ and drains most of the toxins from the matrix or connective tissue via the lymph vessels, which ultimately drain into the superior vena cava. The lymph system is composed of a myriad of little lymph vessels that then aggregate into larger vessels. These larger vessels are interspersed with aggregations of lymphoid tissue, which are composed of immune-competent cells. These lymph nodes, as the aggregations are called, really function as super detection centers for antigens; sensitized immune cells will migrate here to produce millions of similar clones of that sensitized cell. The migration is called *homing* in immunology, and the multiplication is called *cloning*. The swelling seen in these lymph nodes during an infection arises because of the activation of the immune cascade by these sensitized cells. This will cause inflammation of the lymph node.

A major portion of our immune system is located in these lymph aggregations; in fact, the largest part of our immune system is found in the gut lining's so-called Peyer patches. This is why we can manipulate the whole of the immune system by intervening at the level of the gut lining.

### Physiological Considerations

The lymph system is a slow drainage system, and the propulsion of lymph toward the heart is dependent on a number of factors. First, the lymph vessels have no valves but depend on a sort of negative suction action of the truncal vessels, which is similar to that of an amphibian heart. The lymph flows relatively slowly, at a rate of 1 to 2 mL/min, against a high resistance, whereas the venous flow is rapid, at 2 to 3 L/min, against a low resistance. Two-thirds of the body fluid is located in the intracellular space, and one-third is located in the extracellular space (75% of this is in the interstitial space or connective tissue and 25% circulates as plasma). The lymph and venous flow is responsible for circulating most of the extracellular fluid, and the interstitial fluid, in particular, is drained mainly by the lymph system.

It is clear that the following factors will control the fluid interchange:

- The **oncotic pressure** of the plasma and the lymph is determined by the amount of macromolecules, such as protein, and the content of electrolytes, such as sodium and potassium, in the solution. Solutes exert a certain pressure in any fluid, as they "draw water," so to speak, and the more there is of them in a solution, the higher the pressure. When a patient is protein deficient, through malnutrition or disease, we see that there are not enough macromolecules to keep the fluid in the vascular compartment; the fluid will leak out into the interstitium, resulting in edema. Kidney failure, on the other hand, results in an accumulation of electrolytes and other macromolecules in the interstitium. The result is that the oncotic pressure in the interstitium will exceed that of the plasma; edema will again ensue.
- The **hydrostatic pressure** is a mechanical pressure, which can be considered in terms of a hosepipe connected to an open tap. The smaller the diameter of the hosepipe, the higher the pressure; and the more open the tap, the higher the pressure. If there is an obstruction, such as a kink in the hose, the pressure is higher before the kink and lower after the kink.

---

[13] Lukashev ME, Werb Z. ECM signalling: orchestrating cell behaviour and misbehaviour. *Trends Cell Biol.* 1998;8(11):437–441

Fluid always tends to drain from a high-pressure area to a low-pressure area through a semipermeable membrane, which is represented by the venous capillary or the lymph capillary, until the pressure is equal on both sides. Thus, if there is obstruction in the venous system, similar to a kink in the hose, there will be high pressure in the vein and the body will try to equalize the pressure between the vein and the interstitium; thus, the fluid will also accumulate in the interstitium.

- If, for instance, cardiac failure occurs, we see a **back pressure into the venous system**. In our hosepipe model, this will represent a wide open tap, with more pressure in the venous system and so more fluid in the interstitium.
- Last, if there is an obstruction of the lymph flow, we will see **back pressure in the lymph system** and edema will again be the result. We see this in diseases of the lymph vessels, such as elephantiasis in which the lymph vessel is scarred by a parasite.

**The Lymph System as a Detoxifying Organ**

The lymph system has a special relationship with the matrix. The only way out for toxins that are stored in the matrix is via the lymph system. If the matrix is overloaded with toxins, the lymph system will be overloaded with toxins as well because the system has to remove the toxic debris out of the interstitium; at some point, it will also be clogged with toxins. Thus, the stimulation of lymph flow is one of the most important steps to be achieved when cleaning the matrix.

### *The Lungs*

The mucosa of the respiratory tract plays an important role as a barrier to toxins; the lungs are also one of the main points of excretion of gaseous and volatile drugs, such as anesthetics and even alcohol. Elimination via the lungs is typical for toxins with high volatility (e.g., organic solvents). Gases and vapors with low solubility in blood will be quickly eliminated in this way, whereas toxins with high blood solubility will be eliminated by other routes. If they have sufficient vapor pressure, organic solvents absorbed by the GIT or skin are excreted partially in exhaled air with each passage of blood through the lungs. The breathalyzer devise used to test suspected drink drivers is based on this fact. The concentration of $CO_2$ in exhaled air is in equilibrium with the $CO_2$-hemoglobin blood content. Another example is the radioactive gas radon, which appears in exhaled air because of the decay of radium accumulated in the skeleton.

A number of toxins and bacteria are also secreted in the mucus of the respiratory tract, and expectoration is welcomed and supported.

### *The Mucosal Membranes*

Mucosal membranes form the largest part of our bodies in contact with the outside world; they are accordingly very specialized. A mucosal surface is like a microcosm in itself and is a good example of all the components of the ARS active in one organ. With almost 80% of the immune system forming the mucosal-associated lymphatic tissue (MALT), some hormones having receptors on the mucosal cells, a full complement of nerves mediated by the autonomic nervous system, active lymphatic drainage, and the large complement of ECM, the mucosal membrane is one of the most important regulatory organs. It forms a very selective barrier with the tight junction between the epithelial cell and adhesion molecules, playing an important role in deciding what will enter the body; mucosal surfaces can also let some of the immune cells, such as neutrophils, through the tight junction to ingest toxic material in the lumen. The mucosal membrane further protects against toxins by secreting chloride and other solutes into the lumen, which will osmotically draw water to wash away the offender, a process that we see as diarrhea in the gut.

Thus, the integrity of these surfaces is of major importance in the defense against toxins.

The symbiotic gut bacteria also need to be mentioned for their barrier function. These bacteria form a passive barrier against toxins coming into contact with the epithelial surface. They also contribute to the defense against toxins by producing certain metabolites that will serve as fuel for the gut lining and, as such, will help the mucosal cell maintain the integrity of the tight junction. However, because of the hydrolytic enzymes they produce, they can also contribute to the dangerous enterohepatic circulation previously mentioned.

**Absorption via the Gastrointestinal Tract**

Toxins can be ingested by accidental swallowing, intake of contaminated food and drinks, or swallowing of particles cleared from the respiratory tract.

In the case of toxins that are biotransformed in the liver to less toxic or nontoxic metabolites, ingestion may represent a less dangerous portal of entry. After absorption in the GIT, these toxins will be transported by the portal vein to the liver; in the liver, they can be partially detoxified by biotransformation. The active area for absorption in the intestines is about 100 $m^2$.

Some toxic metal ions exploit transport systems specialized for essential elements: thallium, cobalt, and manganese use the iron system, whereas lead appears to use the calcium system.

Many factors influence the rate of absorption of toxins in various parts of the GIT, including the following:

- Physicochemical properties of the toxins: for example, particle size is important (the smaller the size, the higher the solubility).
- Quantity of food present in the gut (diluting effect).
- Residence time in each part of the GIT (from a few minutes in the mouth to an hour in the stomach to many hours in the intestines).
- The absorption area and absorptive capacity of the epithelium.
- Local pH, which governs the absorption of dissociated toxins; in the acidic pH of the stomach, nondissociated acidic compounds will be more quickly absorbed.
- Peristalsis (the movement of the intestines by muscles) and local blood flow.
- Gastric and intestinal secretions transform toxins into more-or less-soluble products; bile is an emulsifying agent that produces more soluble complexes (hydrotropy).
- Combined exposure to other toxins, which can produce synergistic or antagonistic effects in absorption processes.
- The presence of complexing or chelating agents.
- The action of microflora of the gut—about 1.5 kg made up of 60 different bacterial species that can perform biotransformation of toxins.

When we detoxify and drain, it is imperative to support the actions of the gut and also to support the integrity of the barrier function of the gut.

*The Skin*

The skin, which is in constant contact with the outside world, forms the second largest surface of our body after the mucosa. Apart from the barrier function, it is also a major detoxifying organ and has the same P450 system seen in the liver and glutathione to take care of polycyclic aromatic hydrocarbons. The skin can absorb many substances (such as pesticides and chemicals in cosmetic products) and has to be able to detoxify them. Another important function of the skin is to protect us against the harmful ultraviolet (UV) rays of the sun.

Glutathione and other free-radical scavengers, such as catalase and superoxide dismutase, are important because they scavenge the free radicals formed by exposure to UV rays. Induction of the detoxifying P450 pathway will also generate free radicals, as in the liver, so that in the presence of toxins the skin is more exposed to the effects of free radicals. This leads to immunotoxicity, tissue destruction, skin aging, and cancer. Conversely, UV rays damage the detoxifying ability of the skin (the P450 system); sun-damaged skin is less able to deal with toxins.

Because of its role in detoxification, and because it is one of our most important excretory organs (through sweat and evaporation), the skin often bears the brunt when other detoxifying organs, such as the liver, are overloaded. Eczema, drug-induced rashes, and increased sweating are examples. In these cases, it is important to support other organs, such as the liver, for detoxification in skin disease. The liver and skin break down histamine in the body through the P450 system. If the systems are overloaded, allergy will ensue. Histamine and other amines are also formed during the inflammatory process, and many environmental toxins, such as alcoholic drinks, especially red wine, can contain a large amount of histamine.

**Sweat**

Many nonelectrolytes can be partially eliminated via skin by sweat: ethyl alcohol, acetone, phenols, carbon disulfide, and chlorinated hydrocarbons.

**Hair**

Analysis of hair can be used as an indicator of homeostasis of some physiological substances. Exposure to some toxins, especially heavy metals, can be evaluated by this kind of bioassay.

*Other Routes of Elimination*

**Milk**

Many metals and organic solvents and some organochlorine pesticides (DDT) are secreted via the mammary gland in mother's milk. This pathway can represent a danger for nursing infants.

**Saliva**

Some drugs and metallic ions can be excreted through the mucosa of the mouth by saliva (e.g., lead ["lead line"], mercury, arsenic, copper, bromides, iodides, ethyl alcohol, and alkaloids). The toxins are then swallowed, reaching the GIT, where they can be reabsorbed or eliminated in feces.

*Conclusion*

- Many organs are involved in the storage, metabolism, and elimination of toxins once the toxin is absorbed into the body.
- The liver is one of the major detoxification organs and plays an especially important role in the processing of toxins that are absorbed via the oral route.
- The liver is well perfused and equipped with enzymes to render fat-soluble toxins into water-soluble toxins, to be excreted in the kidney and bile.
- The kidney deals with the previously mentioned water-soluble toxins but is also the major organ of elimination, dealing with heavy metals and several drugs.
- The lung plays a major role in the elimination of volatile gases (e.g., organic solvents and gases and vapors with high solubility in the blood).
- The mucous membranes act as a barrier but also contain the P450 system of enzymes and can actively metabolize and excrete toxins.
- The symbiotic microflora play an especially important role here.
- Sweat, hair, milk, and saliva are minor elimination organs and can be used to test the elimination of toxins, such as heavy metals and other toxins.
- Excretion of toxins in milk poses a risk to the infant.
- The matrix and the adipose tissue form a major site of deposition of both water-soluble and fat-soluble toxins. This is a slow-exchange compartment and is in close association with the cell, which is a major area of concern in chronic intoxication.
- The lymph system is the only significant way that toxins can be drained from the matrix and the adipose tissue and, therefore, needs special attention during the drainage process.

### *Why Does a Person Need to Detoxify and Eliminate?*

Toxins stored in the body or not eliminated will be detrimental for various reasons. Toxins can have a wide range of effects, such as fatigue, brain fog, and concentration loss. Other not so apparent manifestations include the so-called chloracne that is caused by halogenated toxins. Some toxins can be endocrine disruptors, can cause immune dysfunction, and can act as carcinogenic substances.

Because of the wide distribution of toxins, our fast lifestyles with modern poor nutrition and toxic food, and the increase in psychological stress, the need for detoxification and drainage exists in every patient. Detoxification and drainage requirements will be different in different populations, however.

The aims of detoxification can be summarized as the so-called "4-S" treatment:

- **S**TOP
  external supply of toxins.
- **S**UPPORT
  the organs of detoxification and drainage.

- STIMULATE
  elimination of toxins.
- SENSITIZE
  the patient for further detoxification.

We saw earlier that toxins have basic kinetics. Toxins need to diffuse across several membranes to reach different compartments and most toxins reach another compartment by passive diffusion across semipermeable membranes. Toxins are carried to and from compartments in the blood and consequently we would like to reduce the concentration of toxins in the blood so that the toxins can start to diffuse back into the bloodstream.

For this reason, the patient is given a nontoxic diet and much fluid during the detoxification period. The supply of toxins, such as inhalants and alcohol, is also proactively stopped. This is the first S: **stop**.

In other cases, the toxins are bound to proteins and to SH groups in the cell and the matrix. We often then have to **stimulate** the release of the toxin from these molecules. This is an active process and needs **support**. To get the body to free itself of toxins, we need to do three things: (1) support the organs that metabolize harmful substances; (2) support the function of the organs, such as the matrix, that store toxins; and (3) stimulate elimination from these organs.

Once stored toxins are released, they often have not completed their metabolism and, therefore, still need to be made water-soluble in the liver before being excreted via the kidney and other organs. If the stored toxins are released too rapidly all at once, or the liver and other metabolizing and elimination organs are overloaded or not functioning properly, the released toxins will diffuse into the blood but cannot be excreted. Thus, they will circulate in the bloodstream until they find a compartment where the concentration is less than in the blood and then diffuse into that compartment. The crux is that toxins are merely shifted from one point to another. This is not such a problem in well persons or in patients with mild toxic effects; however, in patients with severe toxic effects, there may be repercussions, such as heavy metals entering the brain, from where it is extremely difficult to remove them.

This factor needs to be considered, especially in patients in whom the organs of elimination are not functioning properly or who are burdened by disease or other toxins (such as patients receiving chemotherapy). In these patients, we need to support the organs of detoxification and elimination first before we actually drain the tissues.

The process of detoxification and drainage puts a severe burden on the body. Thus, in very frail and sick patients, it can put another burden on the body; in these patients, detoxification should be done as a later procedure, when the patient has received other medications to support the body. Detoxification and drainage also require energy and, therefore, catalysts are a standard addition to more strenuous detoxification programs, apart from the fact that they also play a role in cellular detoxification.

## Immunomodulation

As previously mentioned, most interactions in the human body go through ARSs. Because ARSs use input from other systems and use feedback loops to correct them around a set point, we talk about regulation therapy when we intervene in such ARSs. The immune defense system is also an ARS, which tries to create a purposeful defense against homotoxins of all kind. To achieve this, the initial defensive reaction mobilized in response to contact with the homotoxin will be used as on input to the "own" system to exert control over the level of action of the system against the stressor. Regulation takes place through the release of stimulating and inhibiting mediators, such as cytokines.

The immune system is an example of a subtle ARS in which we can intervene with medications. The therapeutic use of this intervention is called "immunomodulation" when we use medications to affect the immune system. This definition of immunomodulation refers to the action of the medication on autoregulating processes that steer the immunological defense system. Many antihomotoxic medications intervene here; in addition to proven therapeutic efficacy, they are extremely safe.

The reasons for the tolerability of the bioregulatory medicines in antihomotoxic therapy can be found not only in the low concentration of the ingredients, but also in the multitarget intervention that was described earlier (i.e., the so-called Bürgi effect). Complexity is the hallmark of the immune system, and immune processes are characterized by several events that take place either in cascades or at the same time. Linear intervention, as is seen with most of contemporary immune treatments (e.g., NSAIDs), is thus bound to have undesirable effects, some of which cannot be seen in conventional laboratory research because testing is done in isolated systems.

To become more familiar with these complex interactions, it is important to look at the basic actors involved in such processes.

### *Basic Actors of the Adaptive Immune System*

The mammalian immune system has two arms: a nonspecific arm and a specific (adaptive) arm. The latter is found only in higher animals and implies that the immune system has a memory for previously encountered antigens. It also can respond differentially, depending on which antigen is encountered.

These functions are mediated by T-cells, which are a type of lymphocyte. Lymphocytes are divided into B-cells, T-cells, and large granular lymphocytes (LGLs). T-cells are further divided into several classes distinguished by the markers the cells carry on their cell surface. These markers are termed CD (cluster designation) markers and will also give an indication of the function of the cells. The CD4 cells are probably the most familiar, being the cell that is targeted by the human immunodeficiency virus. CD4 cells are helper cells that are activated in all sorts of immune situations to send messengers to the whole immune armamentarium to elicit an appropriate immune response. The CD marker is another name for the TCR (T-cell receptor).

CD8 cells are also composed of various classes. Some are T-cytotoxic cells, some are suppressor cells, and so on. Both the CD4 and CD8 cells need to be shown the antigen via so-called antigen-presenting cells (APCs); these range from the macrophages to the epithelial cells of the mucosal surfaces. A special type of T-cells, the LGLs, include the NK-cell. They do not need an APC but will recognize antigens like cancer cells or cells infected with virus and destroy them.

A lymphocyte may change its surface markers when it becomes mature or when activated, and a host of CD markers have been described that also gives a clue to the function of a particular cell. For instance, T-cells in a neonate will carry the CD marker CD45RA+, a sign that they have not yet encountered an antigen. As soon as they encounter an antigen, they will change that marker in the maturation process. A CD44 cell, for instance, possesses homing receptors that will enable these cells to exit into the peripheral lymphoid tissue, where they will function as mature helper and cytotoxic T-cells[14]. The CD4+ cells can be divided into specific subclasses, such as $T_H1$ or $T_H2$, depending on their functions.

For the adaptive immune response to unfold, three signals are needed[15]:

- Signal 1 derives from the interaction of antigens, in the form of so-called pathogen-associated molecular patterns or plant or organ amino acids with the T-cell receptors on the APCs in association with so-called toll-like receptors. Dendritic cells are specialized APCs and play a central role in stimulating and modulating cell-mediated responses. Infections have a profound effect on dendritic cells via toll-like receptors, which, in turn, interact with T-cells and determine whether $T_H1$- or $T_H2$-type responses develop.
- Signal 2 consists of the activation of co-stimulatory signals.
- Signal 3 is the polarizing signal given by specific cytokines or immune messengers.

**APCs:** These cells will use phagocytosis to engulf antigens to process and present them. Characteristic proteins of the antigen will be presented at the external wall of the APC in a changed MHC class I or II molecule to chemotactically attract T-lympho-

---

14 Roitt I, Brotoff J, Male D. *Immunology.* 5th ed. London: Mosby International; 1998

15 Netea MG et al. From the TH-1/TH-2 paradigm towards a Toll like receptor/T-helper bias. *Antimicrob Agents Chemother.* 2005;49(10)3991–3996

cytes (naive prolymphocytes or $T_H0$) or LGLs. The antigen characteristic (antigen peptide) will be presented from the MHC of the APC to the TCR. If the presentation is via an MHC class II molecule, from that moment on the T-cell becomes a helper cell ($T_H$ cell) that will focus on the antigen defense and stimulate other actors of the defense system to eliminate the invader. Because the triggering by the motif or pattern on the TCR of the helper cell is specific to the function of the antigen, the tasks of the cells are to be seen purely as antigen-specific defense.

The CD4 cells can be subdivided according to the type of cytokines (messengers) they produce.

**$T_H0$ cells:** These cells are naive prolymphocytes; they do not yet have a specific function. $T_H0$ cells can differentiate to proinflammatory $T_H1$, $T_H2$, or $T_H17$ cells or to inflammation down-regulating $T_H3$ cells, also called regulator or $T_{reg}$ cells.

**$T_H1$ and $T_H2$ cells:** These cells are proinflammatory lymphocytes. Their main task is to trigger and stimulate the defense against a specific antigen. They belong to the class of the CD4(+) cells previously mentioned.

$T_H1$ cells are responsible for cellular immunity. They trigger (stimulate) NK cells, macrophages, and even deviating cells from a specific origin (e.g., viral-infected organ cells). They play a major role in defense against viruses and fungal infections. $T_H1$ cells will produce TNF-α, IFN-γ, and IL-2. These cytokines are associated with the inflammatory-type cell-mediated immune responses; they also activate cytotoxic, inflammatory, and delayed hypersensitivity reactions.

$T_H2$ cells direct humoral immunity and are also involved in defense against helminths. Once triggered by a motif on their TCR, their activity results in a stimulation of antibody production (B-cells transform into plasma cells) so that antigens outside the cells (*humoral* = body liquids) are eliminated. $T_H2$ cells produce IL-4, IL-5, IL-6, IL-9, IL-10, and IL-13. This cytokine profile will promote the production of antibody, especially IgE; it therefore plays a major role in defense against helminths and a major role in allergy.

The various cytokines of $T_H1$ and $T_H2$ will inhibit each other and are therefore part of a regulatory process. Depending on the situation, either a $T_H1$ or a $T_H2$ pattern will emerge. When the work is done, however, and the invader is eliminated, the normal $T_H1/T_H2$ balance is restored (**Fig. 4.9**).

**$T_H3$ or $T_{reg}$ cells:** These are inflammation-regulating cells. They inhibit the function of $T_H1$ and $T_H2$ cells and are, therefore, down-regulating in their activity. They belong to the class of CD4(+) cells and produce the cytokine TGF-β, which, in most circumstances, is a down-regulator of inflammation.

**$T_H17$ cells:** Recently, a new type of CD4(+) cell has been added to the previously described repertoire, namely, the $T_H17$ cell. These cells are recruited by IL-6 and TGF-β. It seems that in an environment where there is IL-6, TGF-β will dictate the generation of $T_H17$ cells. These cells secrete mainly IL-17 and seem to have a function against tissue-infiltrating antigens[16]. They may also play a role in autoimmunity. The role of these cells is not clear at this time, but they seem to play a role in diseases in which the $T_H1/T_H2$ paradigm cannot explain all the pathology.

The factors that may influence differentiation into one of these types of cells include the following:

- The type of antigen. For example, if the invader is a virus or a fungus, a $T_H1$ profile will be evoked because a cytotoxic-type response is needed with IFN-γ as an effector. The IFN-γ will further induce IL-2, which will change $T_H2$ cells into $T_H1$ cells, thereby reinforcing the cellular response.
- The amount of antigen present[17].
- The APC and the cytokines they produce (e.g., macrophages bearing an allergen) will invoke a $T_H2$ profile.
- Host genetic background. Half of the offspring of allergic parents will be atopic. It has now been established that patients who are allergic from birth will often be in a $T_H2$–predominant state.

---

[16] Korn T, Oukka M, Kuchroo, V. Betelli E TH17 cells: fffector T-cells with inflammatory properties. *Semin Immunol.* 2007;16:362–371

[17] Ruedl C, Bachmann MF, Kopf M. The antigen dose determines T helper subset development by regulation of CD40 ligand. *Eur J Immunol.* 2000;30(7): 2056–2064

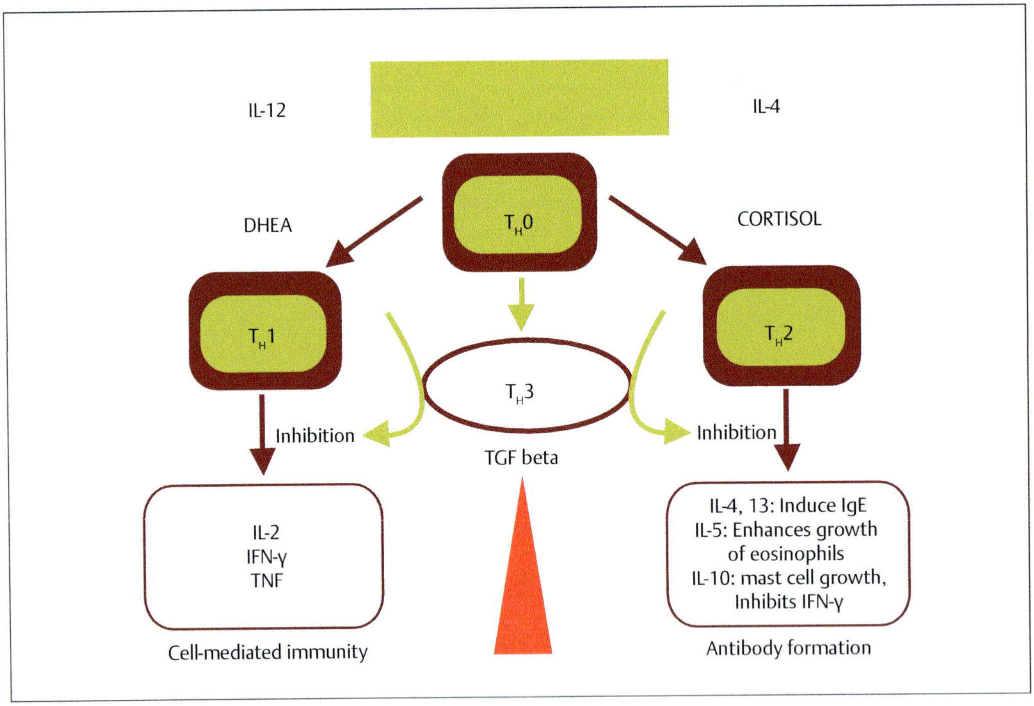


**Fig. 4.9** $T_H1 / T_H2$ balance.

- Activity of the co-stimulatory molecules and hormones in the local environment[18]. This is of particular interest in regulation therapy because the stress hormone cortisol will, for instance, drive the balance toward a $T_H2$ state, whereas its counteracting hormone dehydroepiandrosterone will drive a $T_H1$ response on a tissue level. This explains the effect of stress on certain immune conditions, such as allergy and asthma[19,20,21].

Another important issue is the effects of chemicals[22,23] certain antigens, such as *Candida albicans*; helminths; and parasites. It is well known that parasites can suddenly induce an allergy. Parasites such as bilharzia will produce $T_H2$ predominance. The same "escape" mechanism has been seen in *C. albicans*, which, by virtue of the mannan in its cell wall, will drive the immune system away from the immune state that offers protection against it, namely, the cellular immunity afforded by the $T_H1$ cell. *Candida* can, thus, be a powerful bystander in

18 Hassig A, Kremer H, Liang WX, Stampfli K. The role of the Th1–Th2 shift of the cytokine profiles of CD 4 helper cells in the pathogenesis of autoimmune and hypercatabolic diseases. *Med Hypotheses* 1998 Jul;5 (1):59–63

19 Elenkov IJ, Chrousos GP. Stress hormones, proinflammatory and anti-inflammatory cytokines, and autoimmunity. *Ann NY Acad Sci.* 2002 Jun; 966:290–303

20 Elenkov, IJ. Glucocorticoids and the TH-1 TH2 balance. *Ann NY Acad Sci.* 2004;1204:138–146

21 Hoglund CO, Axen J, Kemi C, et al. Changes in immune regulation in response to examination stress in atopic and healthy individuals. Clin Exp Allergy. 2006 Aug;36(8): 982–992

22 Albers R et al. Distinct immunomodulation by autoimmunogenic xenobiotics in susceptible and resistant mice. *Toxicol Appl Pharmacol.* 1999 Oct 15; 160 (2):156–162

23 Crinnion WJ. Environmental medicine, Part 1: The human burden of environmental toxins and their common health effects. *Altern Med Rev.* 2005;5 (1):52–63

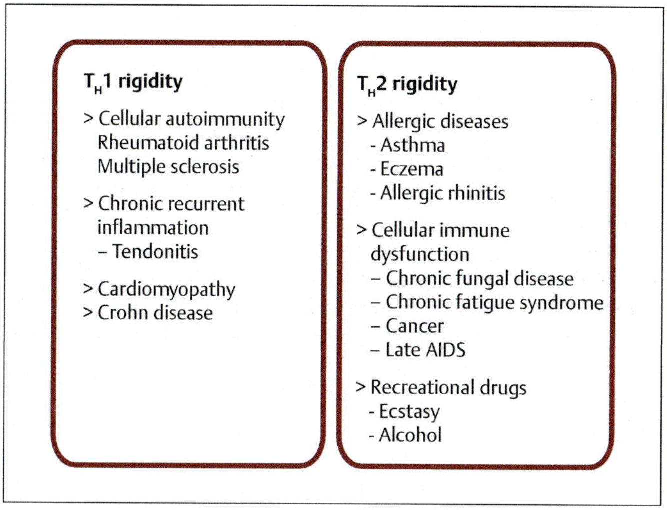


**Fig. 4.10** $T_H1$ / $T_H2$ rigidity.

disease such as allergy, asthma, and even infections[24,25].

As mentioned earlier, because of certain factors, the immune response may lose the ability to modulate between $T_H1$ and $T_H2$, resulting in homotoxicological regulation rigidity. This means that the patient will be predominantly in a $T_H2$ or a $T_H1$ response. This will contribute to certain immune pathologies. A host of diseases can be classified according to their $T_H1$ / $T_H2$ balance (see **Fig. 4.10**). Eosinophilic rhinosinusitis is such a syndrome; the shift into a predominantly $T_H2$ state will predispose the patient toward eosinophilia (the result of IL-4 secretion) and will promote fungal infection. The treatment of such a syndrome is, thus, essentially that of immunomodulation rather than suppression of the fungus or the allergy[26].

## *Applied Homotoxicology and Immunomodulation*

### *Immune Rigidity*

We started this book by mentioning the subtle ARSs in the body that steer the organism towards homeostasis. Homeostasis is a subtle balance in humans (who are open systems) and can, therefore, easily be disturbed by external influences.

The normal oscillation between $T_H1$ and $T_H2$ is of primary importance because this is a normal regulatory phenomenon. The immune system, like the hormonal system, has a chronobiology; it follows that of the hormonal system closely and is affected by sleep pattern[27,28]. Peak production of the proinflammatory cytokines IFN-γ, TNF-α, IL-1, and IL-2 occurs during the night and early morning when the secretion of cortisol is the lowest. Therefore, the body is in a relative $T_H1$ state. The other hormones, such as melatonin and dehydroepiandrosterone sulfate, may also play a role, as they will further stimulate a $T_H1$ response. There is

---

[24] Pucetti P, Romani L, Bistoni F. A Th1 / Th2-like switch in candidiasis: new perspectives for therapy. *Trends Microbiol.* 1995 Jun;3(6) 237–240

[25] Nelson RD, Shibata N, Podzorski RP, Herron MJ. Candida Mannan: chemistry, suppression of cell mediated immunity, and possible mechanisms of action. *Clin Microbiol Rev.* 1991Jan; 4(1):1–19

[26] Lilic D, Cant Aj, Abinum M, Calvert JE, Speckett Gp. Chronic mucucutaneous candidiasis. I. Altered antigen-stimulated IL-2, IL-4, IL-6 and interferon gamma production. *Clin Exp Immunol.* 1996 Aug;105 (2):205–212

[27] Haus E. Chronobiology in the endocrine system. *Adv Drug Deliv Rev.* 2007 Aug 31;59(9–10): 985–1014

[28] Dimitrov S, Lange T, Tieken S, Fehm HL, Born J Sleep associated regulation of T helper 1/T helper 2 cytokinebalance in humans. *Brain Behav Immun.* 2004 Jul;18(4):341–348

even evidence that the $T_H2$ receptor may be blocked at certain points during the night[29,30].

A cell that is gaining more and more importance is the $T_H3$ cell. This is one of the most strongly regulatory cells as far as $T_H1/T_H2$ balance is concerned (i.e., it literally acts as the base of the scale for $T_H1/T_H2$). The $T_H3$ cell secretes TGF-βand occurs mostly in the gut. This is important when immunomodulation occurs via the gut lining.

Heine and others[31–33] recently postulated that homotoxicological products may have their effect on the immune system via the so-called *immunological bystander* reaction. A schematic representation of the immunological bystander reaction can be seen in **Fig. 4.11**. In short, it implies that innocuous plant materials such as *Arnica* and *Bellis perennis* in a preparation like Traumeel, especially in the dilution between D 1 and D 14, will act as an antigen. When Traumeel is ingested, for instance, the APCs of the GALT, notably the mucosal cells and the dendritic cells, will take up the small antigen and present it via the MHC to the $T_H3$ cells. This is so-called motif formation. There will be a different motif for each plant material. The $T_H3$ cells will activate against each of the motifs; thus, there will be one clone of $T_H3$ cells for *Arnica*, one for *Calendula*, one for *Hamamelis*, etc. These $T_H3$ cells will migrate to the nearest lymph node. This is called homing in immunology. In the lymph node, millions of clones of each of the corresponding $T_H3$ cells will be made. They will then migrate to the point of inflammation, where they will balance the $T_H1/T_H2$ oscillations and down-regulate inflammatory phenomena; they can perform these actions because of the secretion of TGF-β.

This is in keeping with the original writings of Reckeweg on homotoxicology, in which he explains the working mechanism of homotoxicological medications. A homotoxin (in this case, inflammatory mediators) will be neutralized when a homeopathic, nontoxic substance mobilizes the innate regulation mechanism of the body[34].

In general, we would also apply different medications for $T_H1$ vs. $T_H2$ rigidity. In $T_H2$ rigidity, we expect the matrix to be rigid as well because this state is often associated with a change in cortisol rhythm. As a rule, Traumeel and Lymphomyosot are used for $T_H1$ rigidity. In the case of $T_H2$ rigidity, more matrix support is needed; products such as Engystol, which has been shown to increase the IFN-γ level and, therefore, will support a $T_H1$ profile, are also used. See **Fig. 4.12**.

A number of substances are being researched in terms of their ability to modulate between $T_H1$ and $T_H2$. The tuberculosis vaccine is such a substance but it can cause severe local skin reactions. This is available in homeopathic form as Psorinoheel or Bacillinum Injeel.

As discussed in the Chapter 1, none of the ARSs of the body will function on their own. Thus, if we want to treat the immune system, we must consider all the other components of it, notably the neuroendocrine system, the liver and gut, the neural reflexes, the matrix, and cellular respiration. In patients with immune dysfunction, more than one of these (or in severe cases, all of them) will be affected. The nutrient status of the patient is of utmost importance; the building blocks for certain vital support cofactors to the whole regulatory process may be lacking.

The gut lining deserves special mention because 60–70% of our immune complement resides in the so-called GALT. The gut mucosa is a highly evolved entity and a study in psycho-neuro-endocrine immunology. It has receptors for hormones such as CRH, ACTH, and thyrotropin, and functions as an immune organ and a selective detoxifier.

---

29 Petrovsky N, Harrson LC. Diurnal rhythmicity of human cytokine production: a dynamic disequilibrium in T helper cell type 1/T-helper cell type 2 balance? *J Immunol.* 1997 Jun;158(11):5163–5168

30 Petrovsky N, Harrison LC. The chronobiology of human cytokine production. *Int Rev Immunol.* 1998;16(5–6):635–649

31 Heine H, Andra F. Zum Antiinflammatorische Wirkungsmechanismus eines Antihomotoxikum compositum. *Aertzteneizeitschrift fuer Naturheilverfahren.* 2002;43(2):96–104

32 Schmolz M, Heine H. Homeopathic ingredients of antihomotoxic medicines modulate TGF 1 synthesis in human whole-blood cultures. *Biologische Medizin.* 2001; 2

33 Heine H. *Homotxicology and Ground Regulation System.* 1st English ed. Baden-Baden: Aurelia Verlag; 2000

34 Reckeweg H-H. *Homotoxikologie Ganzheitsschau einer Synthese der Medizin.* 6th ed. Baden-Baden: Aurelia Verlag; 1986

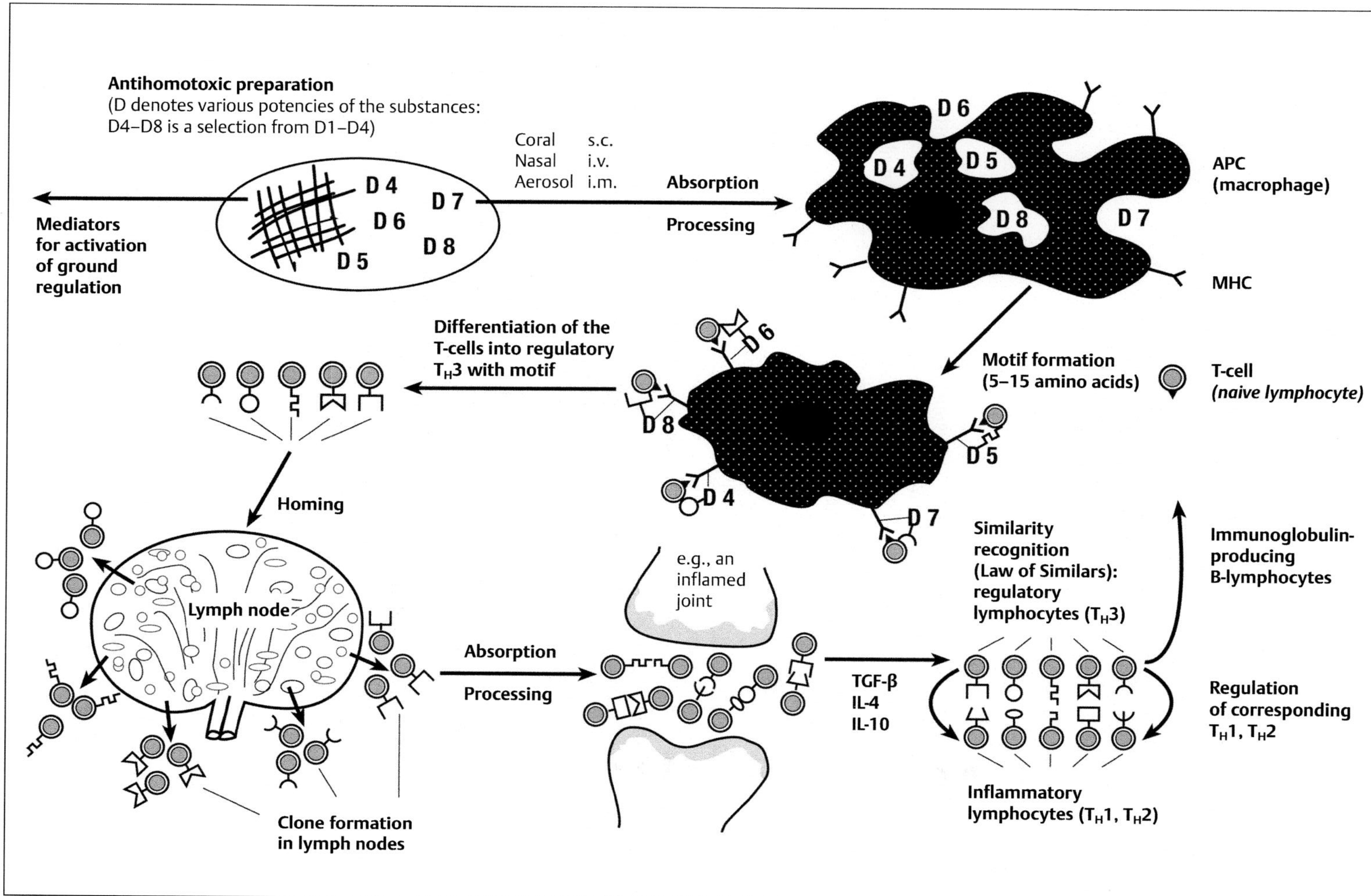


**Fig. 4.11** The immunological bystander reaction.

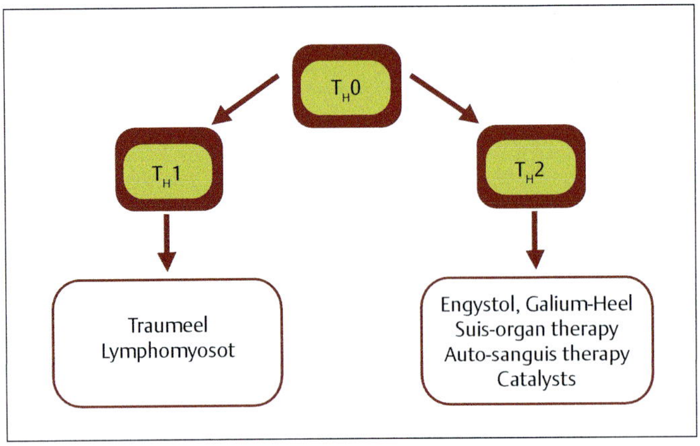


**Fig. 4.12** $T_H1/T_H2$ switch.

Symbiotic bacteria also have a very specific function in immunity[35,36]. By exerting a low-grade antigen load, they are able to induce $T_H3$ cells, apart from the fact that they have a host of other functions, notably contributing to mucosal health. The concept of "mucosal distress syndrome" is also being recognized more and more in modern immunological texts.

Remedies such as Mucosa Compositum and Hepar Compositum are of particular use when treating the gut lining. The presence of the regulatory $T_H3$ cell in the GALT has already been mentioned, and the manipulation of the immune system via the gut lining is becoming more important.

### *Using Complexity in the Immune System*

The immune system is particularly complex because of the myriad of interactions between its own components and between other systems, such as the endocrine and neural systems. In addition, certain very important feedback systems are in place to prevent overreactions of the immune system. Linear interventions disturb this subtle interaction and feedback.

This system can be compared with a mobile hanging from the ceiling. If a person pulls on just one end, the whole system will go out of balance; but they pull on several of the elements at once over the whole system, it will move downward in a balanced fashion. This is basically the way the immune system regulates in nature.

Cortisone is such an efficient immunomodulator because it has a multitarget intervention in the immune process and, therefore, shuts down several processes at once. The problem is the strength and the metabolic side-effects of cortisone. Complex biological substances, such as the plants and animal extracts used in antihomotoxic medicine, have real potential for such multitargeted interventions. From the research cited earlier, the maximal intervention on one target generally does not exceed 70%; accordingly, there may be an effect on several targets. This multitargeted regulation is more often seen in complex homeopathic preparations and in preparations in which plant and animal antigens are mixed. The low concentration in the medication may have a further beneficial effect.

In vitro systems treated with antihomotoxic medications show that both stimulated and understimulated cells can be balanced, thus making these substances true regulators. This also has implications for applied homotoxicology; various opposing immune rigidities can be treated with the same medication without worsening the condition. The complexity is further seen in that the action of the same cytokine may differ in different

---

35 Kalliomaki M, Salminen S et al. Probiotics in primary prevention of atopic disease: a randomized placebo-controlled trial. *Lancet.* 2001 Apr 7;357(9262): 1076–1079

36 Smit A. The leaky gut syndrome. *Biomed Ther.* 1999 Jun;17(3):91–94

circumstances and in different stages of disease. This makes it even more difficult for the common linear intervention seen in conventional treatments. Bioregulatory medicines often support the body's own defense mechanism and, therefore, will allow for all the interactions and feedback systems to remain intact or may only intervene on multiple targets to a certain extent. This gives the body a chance to self-regulate again. An example of multitargeted regulation in an antihomotoxic preparation can be explained with the example of Traumeel.

Traumeel, an inflammation-regulating drug and an excellent immunomodulator, stimulates the production of $T_{reg}$ cells that, by the release of TGF-β, will inhibit the proinflammatory $T_H1$ and $T_H2$ cells[37,38]. The $T_H1$ and $T_H2$ cells stimulate the inflammatory reaction by the release of their proinflammatory mediators at the top of the inflammatory chain in antigen-induced inflammation. Therefore, the immunomodulating action of Traumeel takes place, as far as release of TGF-β is concerned, at the beginning of the process instead of at the end of it; this contrasts with NSAIDs, in which cyclooxygenase inhibition takes place at the bottom of the inflammation cascade. The higher the intervention is in the cascade, the more parameters will be influenced because an earlier-inhibited factor will trigger fewer factors from the top to the bottom of the chain. This is why Traumeel, acting at the top by inhibiting proinflammatory T lymphocytes, will down-regulate all of the following parameters in the inflammation process.

We might even hypothesize that the results obtained in the research described next may, in fact, be a consequence of the TGF-β release at the top of the chain. More recent research has shown Traumeel even to intervene at the level of the secretion of proinflammatory mediators, such as IL-1β, TNF-α, and IL-8[39]. Engystol, an antihomotoxic immunomodulator used mainly in viral infections, has shown an increase of IFN-γrelease of 24% by $T_H1$ lymphocytes[40], down-regulating the $T_H2$ inflammatory pathway (humoral defense pathway). The low concentration of the ingredients and the multitargeted approach to the problem of inflammation will ensure a tolerable, effective immune modulation.

In conclusion, we have in practice today lost the luxury of pure immune suppression or immune activation when dealing with immune syndromes. Most immune syndromes now can be classified as an inability of the immune system to respond appropriately to an antigen. In other words, the immune system has lost the ability to regulate to maintain homeostasis, rather than being merely suppressed or activated.

In homotoxicological terms, we are mostly dealing with regulation rigidity. Homotoxicology lends itself very well to classification and treatment of these syndromes, many of which are newly described, such as chronic fatigue syndrome and chronic eosinophilic fungal rhinosinusitis. We are mostly dealing with a loss of reactivity in our patients. This may be because we have practiced suppressive medicine for so long (i.e., suppressing childhood diseases and the natural innate healing mechanisms of the body). For this reason, we also see an increase in the incidence of diseases such as atopy (allergy), autoimmunity, and cancer[41,42]. By applying bioregulatory treatments, such as antihomotoxic medications, we can effectively restore balance to the immune system.

[37] Heine H et al. Immunologische Beistandsreaktion durch pflanzliche Extrakte in antihomotoxischen Präparaten. *Biologische Medizin.* 1998;27(1):12–14

[38] Heine H et al. The anti-inflammatory action mechanism of an anti-homotoxic composite remedy. *Ärtzezeitschrift für Naturheilverfahren.* 2002;43:2

[39] Porozov S et al. Inhibition of IL-1β and TNF alpha secretion from resting and activated human immunocytes by homeopathic medication Traumeel S. *Clin Dev Immunol.* 2004;11(2):143–149

[40] Enbergs H. Effects of the homeopathic preparation Engystol® on interferon-γ production by human T-lymphocytes. *Immunol Invest.* 2006;35:19–27

[41] Black P. Why is the incidence of allergy and auto immunity increasing? Trends Immunol. 2001 Jul; 22 (7):354–355

[42] Unanue PR. Viral infections and non specific protection. Good or bad? N Engl J Med 2007;357 (13):1345–1446

## Cell and Organ Support

The third pillar within treatment according to the homotoxicological concept in bioregulatory medicine is cell and organ support. Both the cell as a living entity and the organ need our attention.

### Cell Support

*Krebs Cycle Intermediate Products*

Here we analyze the various Krebs cycle intermediate products in detail in order to discover their specific therapeutic role and to learn about their use in ponderal form and in small homotoxicological doses.

From the therapeutic point of view, Krebs cycle intermediates can be used in three different forms with three different groups of often overlapping indications.

**Krebs Cycle Intermediates in Ponderal Form**
This type of administration is substantially supplementary. It is based purely on biochemical and clinical studies with these substances, and laboratory determinations are used.

In practice, if there is an abnormal accumulation of an intermediate in the patient's urine, it is supposed that the enzyme system responsible for its subsequent transformation is blocked. The subsequent metabolite is deficient and is administered in ponderal form.

**Krebs Cycle Intermediates in Homotoxicological Form**
This type of substitution is substantially regulatory. It is based purely on clinical studies and on doctors' experience throughout the world for more than 50 years. Our own clinical experience and that of Reckeweg have contributed to the clinical indications of catalysts in homotoxicological form. In addition, the studies and experience of various naturopaths, particularly Raymund Kastner, must be mentioned. All of these experiments are listed and summarized in Ivo Bianchi's *Homeopathic-Homotoxicological Repertory Materia Medica.*

**Krebs Cycle Intermediates in Homeopathic Dilution**
Krebs cycle catalysts are not distinguished from other homeopathic remedies when used in homeopathic dilution. They are indicated according to experimental symptoms derived from the data of the American homoeopath Dr. David Riley. There is, thus, a true and appropriate *Homeopathic Materia Medica* for many intermediate Krebs cycle catalysts that will not be considered herein, given the strict homotoxicological status of this text.

**Hans Adolf Krebs and Hans Heinrich Reckeweg**
Krebs was born in Hildesheim, Germany, in 1900. His father was a surgeon specializing in otorhinolaryngology. Krebs first studied medicine at the universities in Göttingen and Freiburg, then in Hamburg, where he graduated in 1925. He then studied chemistry in Berlin and was assistant to Otto Warburg at the Kaiser Wilhelm Institute of Biology until 1930. He returned to clinical medicine in Altona, Germany, and subsequently worked in the Medical Clinic of the University of Freiburg, where he performed research and discovered the urea cycle. Being Jewish, he emigrated to the United Kingdom in 1933, where he initially worked at Cambridge with Sir Frederick Gowland Hopkins. Subsequently, he became professor of biochemistry at the University of Sheffield and concentrated his research on intermediate metabolism. In 1937, he discovered the citric acid cycle; and in 1953, he was awarded the Nobel Prize for Physiology for this discovery.

The chronology of Krebs' life and research clearly shows how contact between this scientist and/or his entourage and Reckeweg was very probable. Surely, the founder of homotoxicology breathed the ambience of the great discoveries in the field of biochemistry that took place precisely during the years when he studied and worked in Berlin. This explains the considerable importance attached by Reckeweg to the Krebs cycle intermediate catalysts, which appear in many of the most important homotoxicological drugs. Reckeweg uses these substances with considerable pharmacological skill, demonstrating that his theory is imbued with Krebs' biochemical studies. In homotoxicology, he suggests the use of catalysts in at least four forms:

1. Each catalyst is included in a single dilution, which is generally low, in a complex compound drug that acts on specific organs or tissues. The catalyst is selected in relation to the particular metabolic composition of the specific organ.
2. Each catalyst is provided in various dilutions, in combination, in order to have a rebalancing effect on a specific passage in the Krebs cycle that is altered in relation to the patient's symptoms. There are two possibilities:
   - Blockage of a metabolic reaction, in which case the use of low dilutions of the catalyst itself is mostly, but not solely, helpful; the Injeel forte form of the biochemical substance is indicated.
   - Disturbance, not necessarily in a blocking form, of a metabolic reaction. In this case, a regulatory therapeutic action, which can be obtained with the Injeel form of the catalyst, is useful.
3. The catalysts are proposed all together in combined dilutions or in Injeel form. In this case, the intention is to globally stimulate, reactivate, and rebalance this fundamental metabolic pathway, which can be blocked by wide-ranging factors such as aging, a viral pathological condition, a toxin, or a pharmacological attack. There are several practical methods for the administration of all the intermediate catalysts.
4. The catalysts are proposed all together but only in low dilution, even if differentiated on the basis of the most usual cell metabolism problems. In this case, the catalysts are combined with vitamins, mineral salts, and basic homeopathic remedies, all in a low homeopathic dilution, in order to achieve a direct stimulus of the cellular intermediary metabolism in a simple way that is applicable to practically all patients. These combinations will be useful for children, in whom they act as an anabolic stimulus, and for the elderly, in whom they will support the production of cell energy, which will improve many clinical problems related to metabolic deficit[43].

---

[43] Reckeweg HH. *Homoeopathia Antihomotoxica.* Vol. I. *Eine gesichtete Arzeneimittellehre.* Baden-Baden: Aurelia Verlag; 1991

### Krebs Cycle Intermediate Catalysts in Homotoxicology

These intermediate catalysts are a fundamental contribution of homotoxicology to homeopathy. The introduction of these drugs has allowed practical clinical use of relatively recent acquisitions in cell biochemistry by homoeopathic doctors. We have already hinted at possible, at least intellectual, contacts between Reckeweg and Krebs. Only a deep and direct knowledge of these biochemical problems could have allowed the introduction of various catalysts of the citric acid cycle into homeopathic therapy. Moreover, the excellent function of these catalysts must be considered as striking evidence of the efficacy of the homeopathic preparative method of the drugs.

The Krebs cycle is the common final pathway by which carbohydrates, lipids, and amino acids are oxidized to produce bioavailable energy. A blockage of this complex chain mechanism causes serious cell damage and forces the cell to assume abnormal functioning, such as anaerobic glycolysis, which is typical of a carcinogenic or defective cell and is very extravagant from an energetic viewpoint.

Reckeweg studied complex homotoxicological drugs, such as Coenzyme compositum, with particular care. A true homotoxicological drug for cell regulation should contain all of the various catalysts, coenzymes, vitamins, and trace elements necessary for the proper function of the cell. The dilutions of each substance are selected with considerable care. For example, it is not by chance that Natrium oxalaceticum is present in a lower dilution than other catalysts. In fact, this substance is the critical factor in regulating various points in the cycle. Because the substance is fairly sensitive to metabolic disorders, its mitochondrial concentration is critically low; it therefore needs a larger bioavailable share, which is made possible by the lower dilution in which the substance is present.

Most impregnation phases and, therefore, all degenerative and neoplastic phases are marked by cell aerobic respiration disorders. The administration of citric acid cycle acids or salts stimulates the intermediate functions of the respiratory chain and compensates for current metabolic lesions.

The administration of these catalysts must obviously be very frequent in neoplastic phases; in other cases, they are generally given once or twice a week and even farther apart as the condition improves. If regressive vicariations appear, administration is suspended until the correct identification of the new clinical situation.

We should not forget that the Krebs cycle is the principal source by which—by the formation of ATP—the cell obtains energy[44]. An enzyme blockage at any level causes a blockage in mitochondrial respiration and can be the final cause of its impregnation or degeneration. In a carcinogenic cell, the blockage of the citric acid cycle leads to an accumulation of lactic acid (produced by the anaerobic metabolism of pyruvic acid) and of many other unused intermediate catalysts. The administration of barbiturates has similar metabolic effects on normal cells: the addition of suitable homeopathized catalysts has an unblocking action on cell metabolism and eliminates congestion.

#### Prescriptive Possibilities and Indications of Intermediate Catalysts

1. **Phase structure.** A patient in the impregnating phase always needs cell support with catalysts from this phase on.
2. **Possibility of influencing the Krebs cycle.** The Krebs cycle can be influenced in several ways: unblocking with the individual catalyst or combination, global stimulation and restart with a single package of catalysts, and general support of cell functions with Coenzyme compositum.
3. **Possibility of combining the catalysts with other homotoxicological drugs.** The catalyst is part of the homotoxicological strategy, together with the nosode, the single homeopathic agent, and the organotherapeutic agent. These agents should all be specifically considered for each patient.
4. **One or more catalysts can be prescribed by homeopathic methods.** Catalysts should be prescribed on the basis of the homotoxicological listing of the patient's symptoms.

[44] Mathews CK, van Holde KE. *Biochemistry*. Redwood city, CA: Benjamin/Cummings; 1990

*The Catalysts*

#### Natrium Pyruvicum

##### *Elements of Biochemistry and Physiopathology*

Pyruvic acid is a tricarboxylic ketoacid produced during the final phase of cytoplasmic anaerobic glycolysis. In the mitochondrion, pyruvic acid can be reduced to lactate and released into the cytoplasm, oxidized to acetyl-coenzyme A, and inserted into the Krebs cycle.

Research has shown that taking pyruvate (**Fig. 4.13**), the salt of pyruvic acid, can increase muscle resistance and promote fat loss. To understand this, we must simply realize that if some cells, such as red corpuscles, use only the anaerobic path of energy production, this favors considerable use of pyruvic acid. The muscle cells, even when at rest or during normal stress, mostly use the anaerobic path of energy production, consuming a large amount of glucose under stress, transforming it first into pyruvic acid and then into lactate (**Fig. 4.14**). This explains why physical exercise is very good for the prediabetic or diabetic patient, aside from the aspect of weight reduction, precisely through consumption of glucose and inhibition of insulin peaks. The administration of pyruvic acid activates this metabolic path and, consequently, the amount of glucose taken up by the muscle cell from the bloodstream. This substance, therefore, increases quickly available cell energy and, in addition, favors the accumulation of glycogen in the muscles for future energy requirements. A recent study demonstrated that pyruvic acid increases the muscle uptake of glucose by 300% and the muscle synthesis of glycogens by 50% after 1 hour of physical exercise. The strength of the arms increases by 150%; and of the legs by 60%. A study performed at the University of Pittsburgh in Pennsylvania has shown that pyruvic acid produces a significant loss of weight and fat in obese women on a hypocaloric liquid diet. In order to explain these data, we can refer to the ability of pyruvic acid to increase both the resting metabolism and the use of fat. Much research today indicates that excessive physical exercise is harmful; even this is easily explained on the biochemical level. Excessive exercise stimulates the use of cell glucose and the production of pyruvic acid and lactic acid. This induces tissue acidosis, which has an

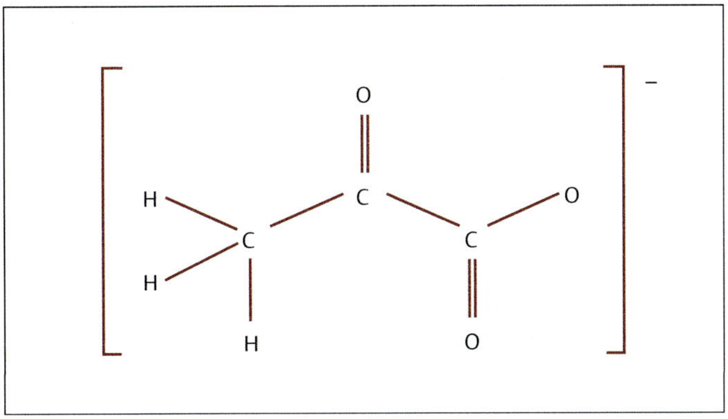


**Fig. 4.13** Pyruvic acid.

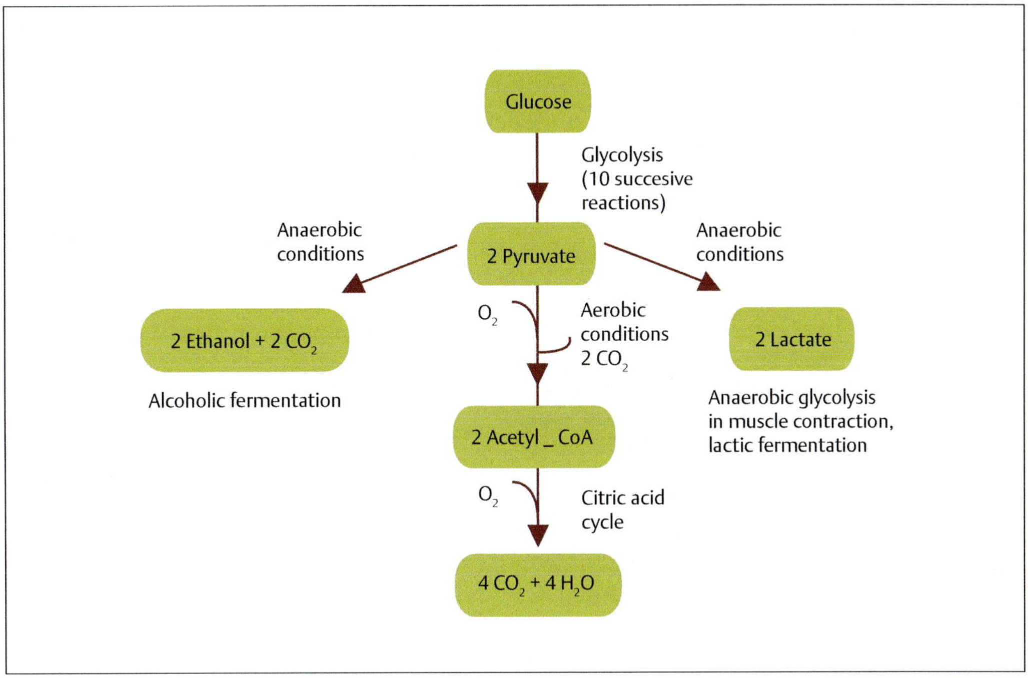


**Fig. 4.14** Pyruvate in anaerobic energy production.

effect of immunological inhibition that sometimes, especially if prolonged, can prove dangerous. Muscle stimulus (such as intense physical exercise) should always be combined with intake of alkalizing substances (such as a vegetable-rich diet) and with suitable tissue drainage. An interesting alternative is the administration of pyruvate in homotoxicological doses, which has a regulatory effect on this metabolic pathway, tending to make it less tumultuous and to ward off excessive formation of lactic acid. On the other hand, in the healthy subject, this metabolism must be accelerated to avoid intracellular accumulations of pyruvic acid, which stimulate anaerobic metabolism (**Fig. 4.14**).

Pyruvic acid, especially in the form of the calcium salt, is widely used in slimming diets because of the stimulus that it exercises on the catabolism of the fatty tissue via the insulin inhibition

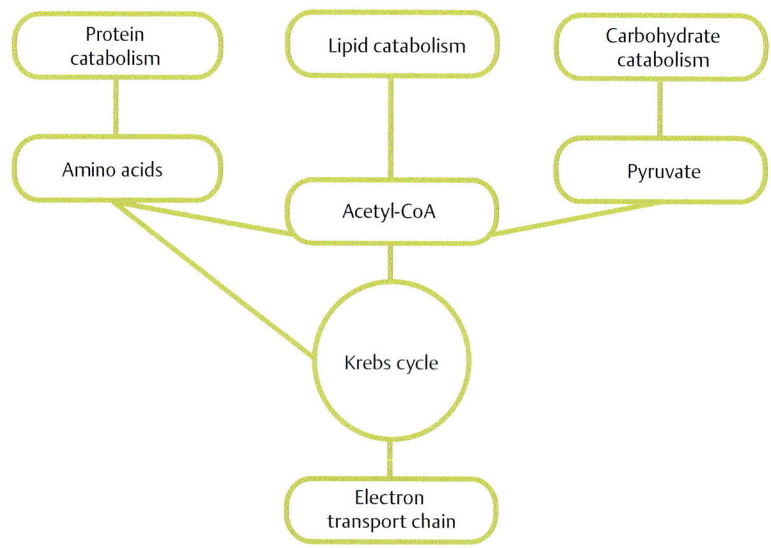


**Fig. 4.15** Pyruvate in lipid metabolism.

described above (**Fig. 4.15**). According to Ronald Stanko of the University of Pittsburgh, called "the father of pyruvate," these diets do not use large quantities of the substance: 5 g / day is sufficient to obtain good metabolic results.[45]

Recent research has confirmed that pyruvate specifically stimulates the loss of fat. In one study, 6 mg / day of pyruvate was administered in combination with 250 µg of chromium for 6 weeks to patients on a diet of 2000 kilocalories (8400 kilojoules) a day. This was attended by a decrease of 12.4 % in body fat mass with a corresponding increase in lean mass. These patients also mentioned an increase in energy and a reduction in tiredness in 71 % of cases[46–50].

[45] Stanko RT, Robertson RJ, Spina RJ, Reilly JJ, Greenawalt KD, Goss FL. Enhancement of arm-exercise endurance capacity with dihydroxyacetone and pyruvate. *J Appl Physiol.* 1990; 68:119–124

[46] Stanko RT, Robertson RJ, Galbreath RW, Reilly JJ, Greenawalt KD, Goss FL. Enhanced leg-exercise endurance with a high-carbohydrate diet and dihydroxyacetone and pyruvate. *J Appl Physiol.* 1990; 69:1651–1656

[47] Stanko RT, Tietze DL, Arch JE. Body composition, energy utilization, and nitrogen matabolism with a 4.25 MJ/d low-energy diet supplemented with pyruvate. *Am J Clin Nutr.* 1992;56:630–635

[48] Stanko RT, Tietze DL, Arch JE. Body composition, energy utilization, and nitrogen matabolism with a severely restricted diet supplemented with dihydroxyacetone and pyruvate. *Am J Clin Nutr.* 1992; 55:771–775

[49] Stanko RT, Adibi SA. Inhibition of lipid acculuation and enhancement of energy expenditure by the addition of pyruvate and dihydroxyacetone to a rat diet. *Metabolism.* 1986;35:182–186

[50] Stanko RT, Ferguson TL, Newman CW, Newman RK. Reduction of carcass fat in swine with dietary addition of pyruvate and dihydroxyacetone. *J Animal Sci.* 1989;67:1272–1278

### Homotoxicological Use

The sodium salt of pyruvic acid, Natrium pyruvicum, is one of the most useful and most used catalysts of the Krebs cycle in homotoxicological form.

It has proved to be particularly useful in tissue toxin drainage, especially when this is correlated with the presence of a chronic metabolic pathological condition such as diabetes, in which case there is a profound disturbance of the metabolism of this substance, or when it is the consequence of an acute infection that has, in some way, blocked the Krebs cycle and, therefore, the use of pyruvate.

Natrium pyruvicum-Injeel activates cytoplasmic anaerobic glycolysis and, consequently, the use of glucose, playing an important role in diabetic patients, particularly when associated with the other catalyst stimulating the use of glucose (DL-malic acid). This catalyst pair is, therefore, specifically therapeutic in type 2 diabetes. If administered parenterally twice a week, this pair antagonizes much of the metabolic harm of diabetes (the primary origin of which is being researched in cell biochemistry). It will always be useful to associate intermediate catalysts with a therapeutic organ, drainage remedies, and a specific homeopathic remedy to effect the synergic therapeutic association that is typical of homotoxicology.

**SENILE DIABETES: Stimulus to the Use of Glucose**

Natrium pyruvicum-Injeel—Acidum DL-Malicum-Injeel—Musculus suis-Injeel—Gallium-Heel—Acidum sulfuricum-Injeel forte
Administered together intramuscularly or intravenously twice a week

If there is correct metabolism of pyruvic acid to acetyl-coenzyme A, then the citric acid cycle can be engaged and energy and heat can be produced, starting from acidotic toxin disposal in the tissue. A good level of energy production also allows the body protection over the nuclear structures that is essential so that the acute illness, often a seriously catabolic and debilitating event in the cell, can be overtaken by a complete "restitutio ad integrum" of the cell and a functional recovery from the biochemical viewpoint.

It has also been noted that Natrium pyruvicum, particularly when associated with Natrium oxalaceticum, has a considerable action on the tissue metabolism of fats. In particular, the subcutaneous injection of the 2 catalysts in Injeel form weekly has proved useful in hepatic steatosis. In this case too, it will be useful to include the appropriate catalysts within a specific therapeutic strategy.

**HEPATIC STEATOSIS: Stimulus of Fat Catabolism**

Natrium pyruvicum-Injeel—Natrium oxalaceticum-Injeel—Hepar suis-Injeel—Phosphorus-Injeel
Administered together subcutaneously over the dermatome of the liver once a week

### General Considerations

Homotoxicologically, the various homeopathic dilutions of Natrium pyruvicum act as a specific remedy for reactive deficit and the tendency to chronicization and progressive vicariation towards more serious and profound pathological forms. Natrium pyruvicum is useful both for giving reactive force to the body after an acute pathology and for antagonizing cell metabolic deficit, starting from a chronic pathological condition, which inevitably occurs with aging[51–53].

## Acidum Citricum

### Elements of Biochemistry and Physiopathology

Isolated for the first time by the German biochemist Karl Wilhelm Steele in 1784, citric acid (**Fig. 4.16**) is used today in many clinical situations, from fatigue to digestive difficulties, from a cold to influenza and asthma, and from hypertension to atherosclerosis. Citric acid is, in various quantities, present in all vegetable and animal tissues. It plays a crucial role in the Krebs cycle, originating from the reaction of oxaloacetate with acetic acid, which is a subunit of acetyl-coenzyme A produced by the degradation of fats, glucose, and amino acids. If correct metabolism of citric acid does not occur, it tends to accumulate and create problems, including abnormal absorption of toxic metals, such as aluminum, involved

[51] Bianchi I. *Argomenti di Omotossicologia*. Vol. 2. Milan: Guna Editore; 1990

[52] Bianchi I. *Argomenti di Omotossicologia*. Vol. 1. Milan: Guna Editore; 1987

[53] Bianchi I. *Principles of Homotoxicology*. Vol. 1. Baden-Baden: Aurelia Verlag; 1987

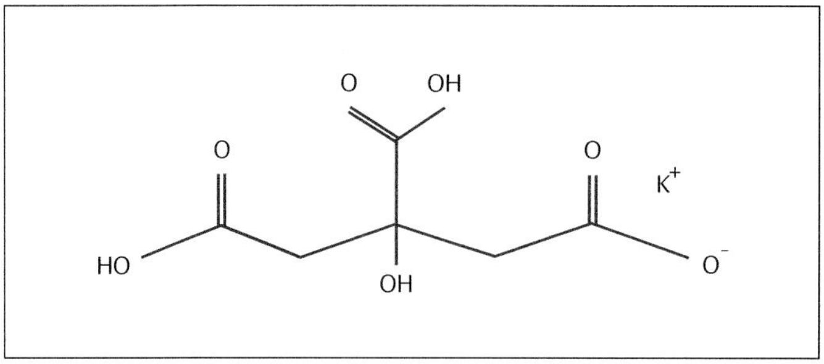


**Fig. 4.16** Acidum citricum.

in the pathogenesis of Alzheimer disease. If citric acid is not correctly metabolized by aconitase and isocitric dehydrogenase, which occurs in the case of deficiency of iron, it accumulates and considerable damage occurs, in particular at the vascular level. The absorption of aluminum favored by the accumulation of citric acid occurs to a larger extent the more iron is deficient because these two elements share the same transport systems (in the absence of iron, transferrin transports aluminum).

We must also not forget that citric acid is at the start of the biosynthetic pathway of fatty acids and cholesterol. Therefore, its accumulation inevitably leads to an enhanced synthesis of these substances and results in a greater risk of atherosclerosis.

As already observed by Krebs himself, the citric acid cycle (**Fig. 4.17**) anions *iso*-citric and *cis*-aconite stimulate both the consumption of oxygen and the oxidation of pyruvate, in a much more pronounced way than expected in relation to the quantities administered, suggesting a catalytic role in the cycle.

Recent studies have demonstrated the considerable role that control of the metabolism of citric acid can play in prostate carcinoma[54]. Although it is true that most malignant tumors derive energy for growth from an increase in anaerobic glycolysis, which justifies their avidity for glucose, prostate carcinoma is distinguished by slow glycolysis and a meager avidity for glucose. On the other hand, it has been confirmed that prostate carcinoma is accompanied by changes in lipid metabolism. This should not be surprising in view of the known hormone dependence of this neoplasm. If we also think about the position of cholesterol in the biosynthetic pathway of the sex hormones, these remarks are absolutely logical (**Fig. 4.18**).

In fact, many enzymes involved in the metabolism of fats are altered in these patients. In particular, the β-oxidative path is stimulated. The increase in the use of fatty acids leads to an increase in both ATP and acetyl-coenzyme A (**Fig. 4.19**). This increased availability accelerates the availability and oxidation of citric acid, which is an important source of energy for this type of neoplastic cell. The blockage or modulation of this metabolic pathway should be considered in patients suffering from prostate carcinoma and in all probability in those suffering from hormone-dependent mammary carcinoma.

### Homotoxicological Use

Acidum citricum is one of the most interesting catalysts in homotoxicological form. Its clinical indications are completely rational on the basis of the preceding considerations. In particular, this preparation has a tropism for the nerve structures that suffer in the event of blockage of the anabolic pathway of the lipids from which they are composed. One specific indication is neurological damage in general, and toxic neurological damage in particular. This frequently occurs, for example, in autistic subjects; these subjects improve with an extremely fat-rich diet. Another specific neurological indication is motor incoordination.

[54] Costello LG, Franklin RB. The intermediary metabolism of the prostate: a key to understanding the pathogenesis and progression of prostate malignancy. *Oncology.* 2000;59: 269–282

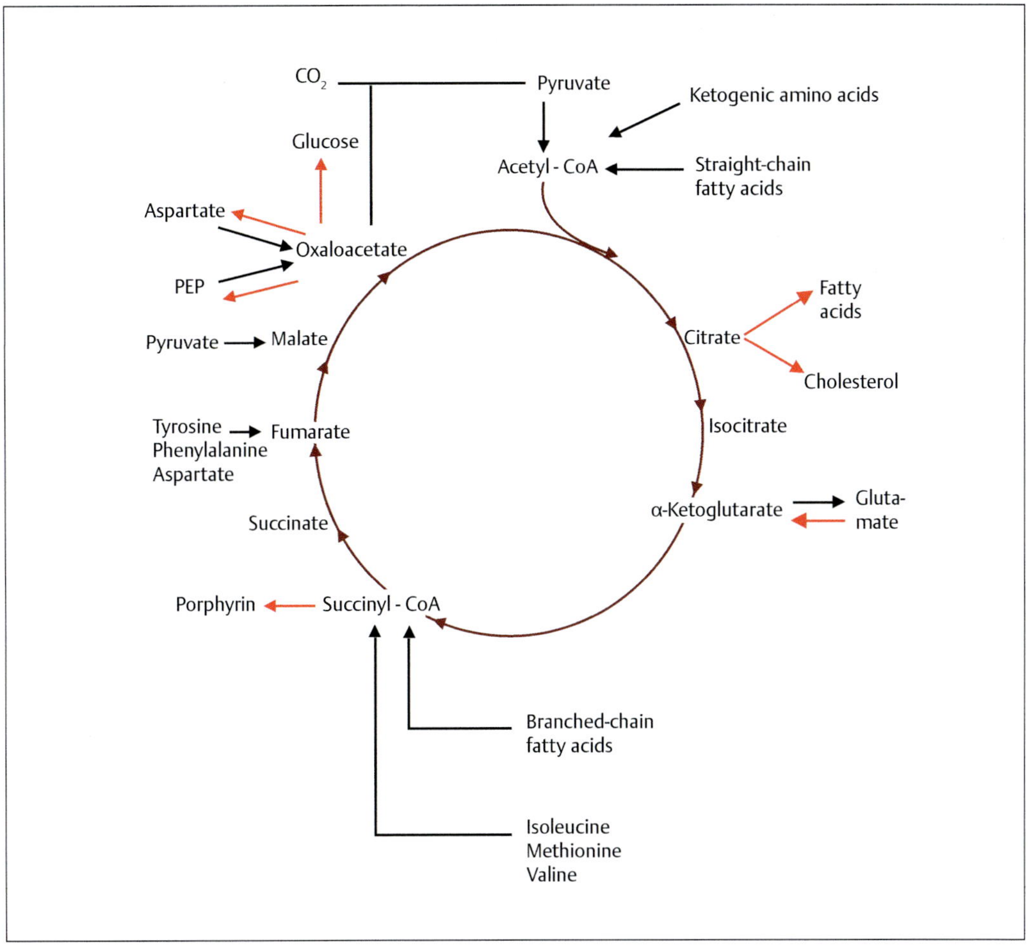


**Fig. 4.17** Citric acid cycle.

***NEUROLOGICAL TOXIC DAMAGE: Stimulus of Nerve Tropism***

Acidum citricum-Injeel—Cerebellum suis-Injeel—Cicuta-(single medication)
Administered together intramuscularly or orally twice a week

In addition, the blockage of the biosynthetic pathway of cholesterol, which occurs when citrate metabolism is disturbed, induces a deficiency of substances such as dehydroepiandrosterone, which are fundamental to the correct functioning of the immune system. This explains why homotoxicological citric acid to stimulate or modulate this metabolic stage of the Krebs cycle is indicated in situations such as lymphatism, weight loss, asthenia, and allergic pathological conditions.

Citric acid favors the absorption and use of calcium. Therefore, the use of Acidum citricum-Injeel, even as a local injection, is extremely important in osteoporotic problems or rickets.

***OSTEOPOROSIS: Stimulus of Calcium Metabolism***

Acidum citricum-Injeel—Os suis-Injeel—Glandula parathyroidea suis-Injeel—Calcium phosphoricum-Injeel
Administered together subcutaneously at the spine or femoral neck once a week

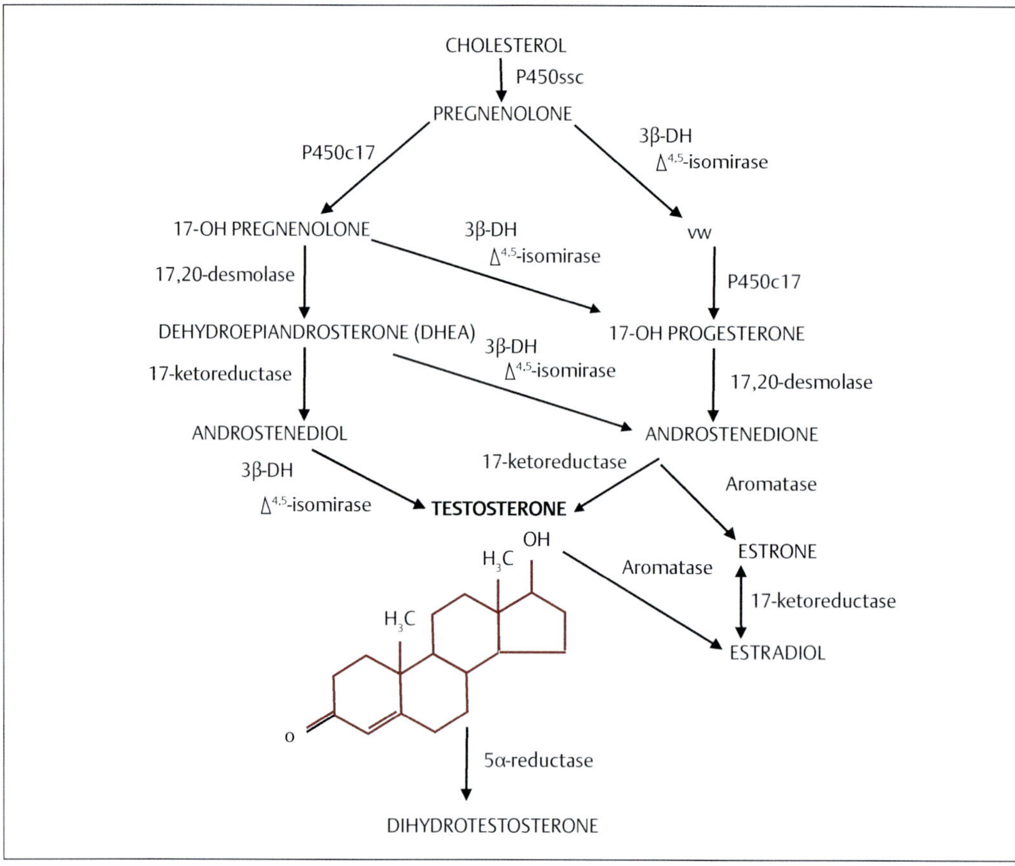


**Fig. 4.18** The position of cholesterol in the biosynthetic pathway of the sex hormones.

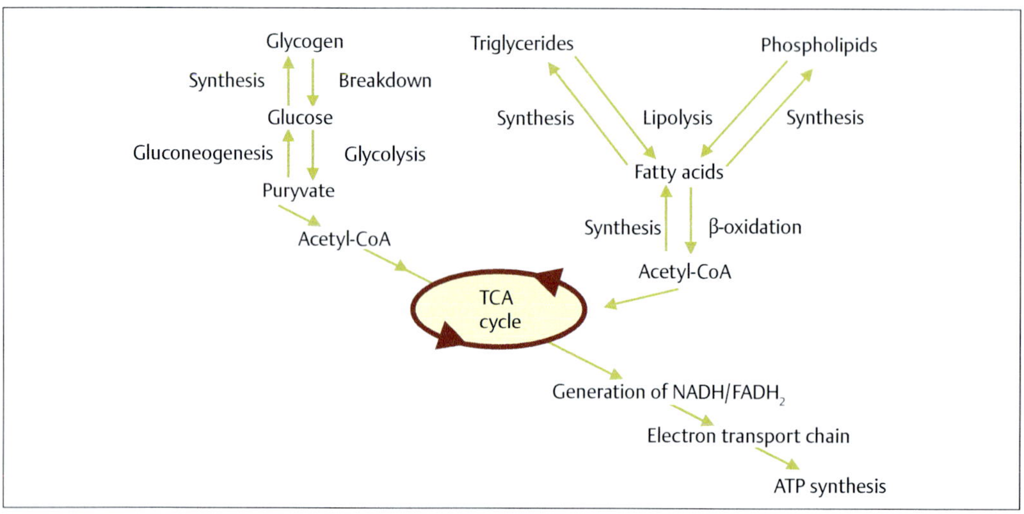


**Fig. 4.19** The increase in the use of fatty acids leads to an increase in both ATP and acetyl-coenzyme A.

### General Considerations

Homotoxicologically, the various homeopathic dilutions of citric acid act as a specific remedy for deficit and hypofunction, particularly useful for stimulating digestive secretions and consequently assimilative processes. Citric acid is typically bimodal: a first stage of hyperfunction and hypersensitivity is followed by the stage of exhaustion and torpor that determines a large part of the indicated pathology. Citric acid is a fundamental cell and tissue activation remedy that intervenes in the metabolism of lipids, and therefore influences physiologically fat-rich tissues, and in the metabolism of calcium, influencing neurological function and bone structure[55].

## Acidum *Cis*-Aconitum

### Elements of Biochemistry and Physiopathology

*Cis*-Aconitic acid (see **Fig. 4.20**) is an intermediate metabolite of the transformation of citric acid into *iso*-citric acid and seems to be an obligatory pathway of the Krebs cycle even if its presence in the tissue is very fleeting.

Iron-sulfur clusters are essential for this type of enzyme reaction. In this case, they do not act as electron transporters, as in oxidative phosphorylase; instead, they interact directly with the enzyme substrate by being converted into the inactive form (**Fig. 4.20**).

*Cis*-Aconitic acid is involved in the metabolism of leucine and influences the degradation of messenger RNA in the cytoplasm.

In chronic fatigue syndrome, a high level of cell *cis*-aconitic acid has been shown to indicate poor functioning of the enzyme mechanisms in use. In addition, according to several researchers, the intracellular accumulation of this substance induces the accumulation of activated iron-sulfur clusters, which can induce serious oxidative damage in the cell structures by excessive and/or abnormal activation of the mitochondrial phosphorylations. This may explain the results of some recent research that connects the decrease in the aconitase enzyme with aging, so that making the determination of aconitase decarboxylase a true and appropriate marker of physiological aging is being considered. Research on *Drosophila* has demonstrated that there is a direct correlation among levels of oxygen consumption, production of free radicals, and a decrease in the aconitase enzyme, with a consequent accumulation of *cis*-aconitum acid in the cell. Among the various proteins involved in the Krebs cycle, only this enzyme tends to decrease in such a significant manner with age. This is probably because of the particular structure of the protein, which is easily oxidized in an environment otherwise rich in iron-sulfur clusters; these clusters activate the oxidative enzymes. Why this enzyme damage can cause an acceleration in the aging processes is easily explained when we consider that it leads not only to a global slowing of the Krebs cycle but also to an abnormal accumulation of peroxide radicals[56,57].

### Homotoxicological Use

The homotoxicological form of *cis*-aconitum acid has a notable moderating importance on these biochemical reactions. A whole series of clinical conditions, from hyperactivity to violent allergic or febrile, with at times disproportionate manifestations with respect to the triggering cause, is attenuated by the administration of this preparation. Therefore, this agent can be used in young and in hyperexcitable patients at all levels, for nervous, immunological, and cardiac indications. The stimulation of this reaction, even if delayed when it appears in the physiological process of aging, can be an important component in a well-constructed antiaging homotoxicological therapy.

---

[55] Bianchi I. Chinoni, catalizzatori, suis: gli apporti originali di Reckeweg all'omeopatia. *La Medicina Biologica*. 2005;(4):21–29

[56] Campbell MK, Farrell SO. *Biochimica*. Naples: EdiSES srl; 2004

[57] Yarian CS et al. Aconitase is the main functional target of aging in the citric acid cycle of kidney mitochondria from mice. *Mech Ageing Dev.* 2006;127 (1):79–84

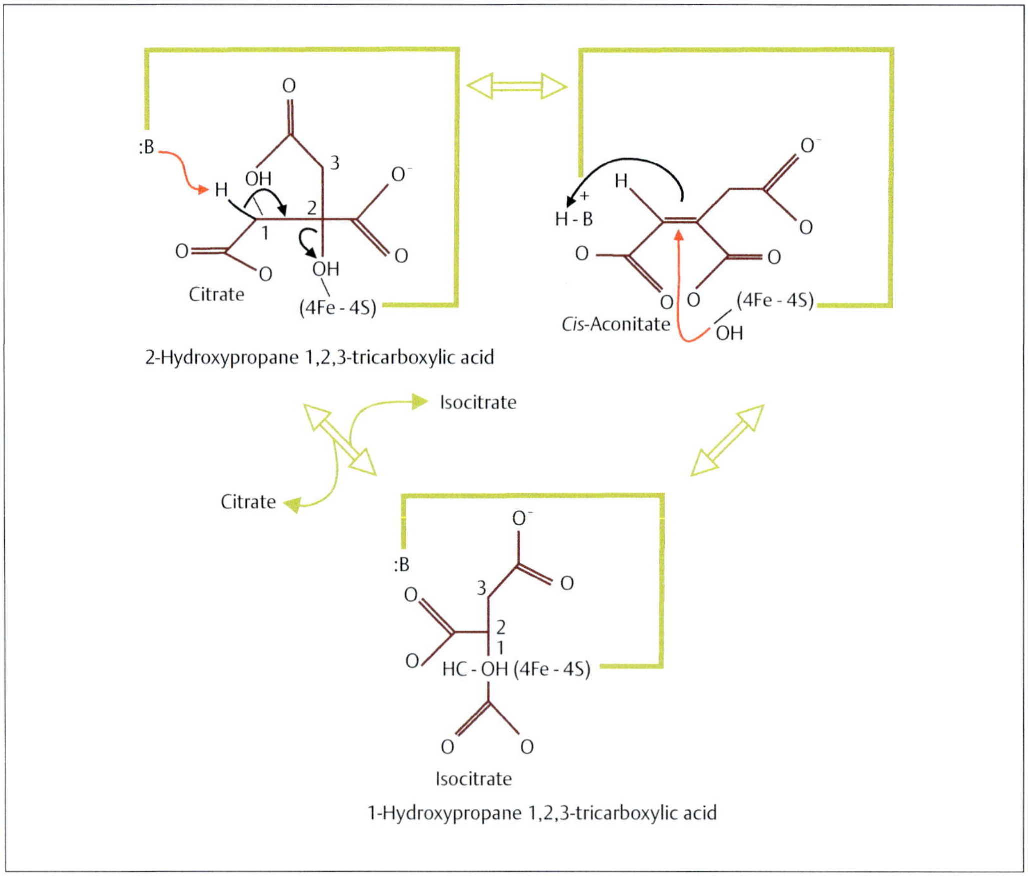


**Fig. 4.20** Aconitase reaction mechanism.

### *General Considerations*

Homotoxicologically, the various homeopathic dilutions of *cis*-aconitum acid act as a specific remedy for nervous hyperexcitability, tendency to somatization, allergies, exaggerated sensitivity to stress, toxins, stimulants, heat, and cold. All the symptoms are violent and unexpected. The symptoms are sometimes intermittent with regard to the episodic presence of hyperstimulating substances in the environment (e.g., pollen, dust, and mold) and the natural periodicity of events that are always connected to the seasons or to the phase of the moon. It is an important drug for the reactive phase whenever this is exaggerated.

## Acidum α-Ketoglutaricum

### *Elements of Biochemistry and Physiopathology*

α-Ketoglutaric acid (**Fig. 4.21**) is a fundamental stage in the Krebs cycle. Its role of precursor is fundamental in the synthesis of glutamic acid, a nonessential amino acid that the body uses for both protein and glucose synthesis.

Glutamic acid has the important function of stabilizing blood glucose levels, particularly during intense physical exercise, fasting, or very low-energy diets. Its supplementation in ponderal doses is important for athletes who want to achieve very high performance levels. Another metabolic role of glutamic acid is that of buffering ammonium derivatives. Ammonium is absorbed in the intestine and transported directly to the liver.

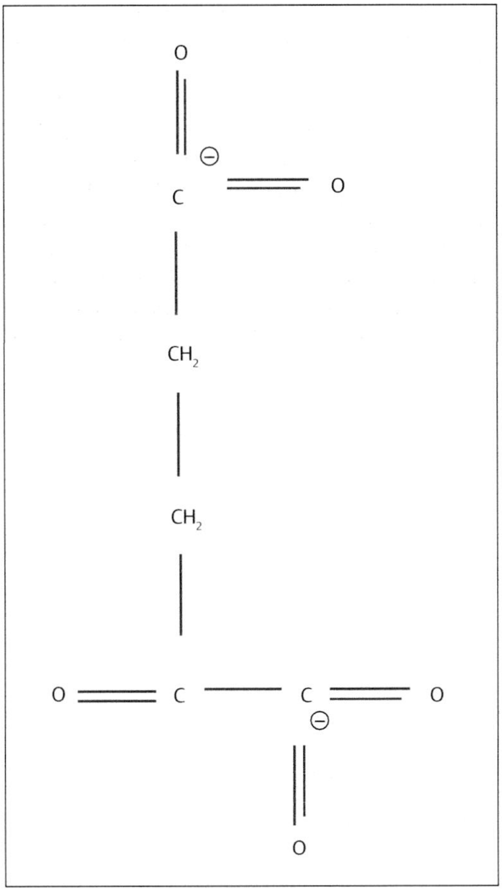


**Fig. 4.21** α-Ketoglutaric acid.

Under normal conditions, the liver extracts the ammonium completely from the bloodstream. However, in the case of excessive intestinal absorption, considerable production in the muscles, or compromised hepatic purifying capacity, ammonium stays in circulation and tends to damage the central nervous tissue. A larger quantity of α-ketoglutaric acid is produced and stocked in the muscles of a trained athlete, with a buffer effect against ammonium. Several studies have demonstrated how ammonium, which is formed in the muscles during physical exercise, can be conjugated and inactivated with the administration of α-ketoglutaric acid and L-glutamate. This allows somewhat more prolonged muscular exertion. The administration of α-ketoglutarate to patients after surgery has resulted in a reduced loss of lean mass and in conservation of nitrogen. For similar reasons, it can be used to increase athletes' muscle mass[58].

### *Homotoxicological Use*

α-Ketoglutaric acid has important applications in homotoxicological form, particularly in a whole series of ailments of the neuromuscular structures. It is above all indicated in asthenia, particularly when, as in chronic fatigue syndrome, the condition is induced by a viral problem that has presumably directly interfered with the metabolism of this substance in the Krebs cycle. Other indications include postural problems, cardiac asthenia, atonic constipation, and alterations of muscle contractility, from incontinence to sphincter spasm. These symptoms can be the consequence of viral infection, iatrogenic intoxications, or stress. In these cases, homotoxicological α-ketoglutaric acid will always be of assistance.

According to the individual case, an intravenous method or a local subcutaneous method must be chosen. The intravenous method results in a general effect, which is necessary for conditions such as chronic fatigue syndrome, whereas the local subcutaneous method results in the treatment of a specific mass or muscle structure. Intramuscular administration will have a more general restorative and trophic effect on the musculature and will be particularly indicated for athletes.

### *General Considerations*

Homotoxicologically, the various homeopathic dilutions of α-ketoglutaric acid act as a specific remedy for hyporeactivity, anergy, and tendency to exhaustion (after a fleeting period of hyperfunction) of the tissue structures, sometimes connected with psychological stress and sometimes after viral ailments[59]. α-Ketoglutaric acid is an

---

[58] Blomqvist BI et al. Glutamine and a-ketoglutarate prevent the decrease in muscle free glutamine concentration and influence protein synthesis after total hip replacement. *Metabolism.* 1995;44(9): 1215–1222

[59] Bianchi I. Incidenza dello stress sui fattori dell'invecchiamento – Terapia ragionata omotossicologica. *La Medicina Biologica.* 2007;(43):25–35

important stimulatory remedy and is very useful in autoimmune diseases and in paucisymptomatic degenerative pathological conditions; in general it can always be used after events of a debilitating nature, which tend to chronicization or at least to residual effects, particularly at the neuromuscular level.

### Acidum Succinicum

#### *Elements of Biochemistry and Physiopathology*

Succinic acid is a key stage in the Krebs cycle because it is the method of direct communication between the cycle and oxidative phosphorylation. The oxidation of succinic acid to fumaric acid requires the reduction of a molecule of flavin–adenine dinucleotide (FAD) from the internal membrane of the mitochondrion, where this is, in turn, oxidized with the intervention of coenzyme Q. Therefore, if an inadequate quantity of coenzyme Q is present, these reactions cannot occur; the Krebs cycle is blocked at this level and an abnormal amount of succinic acid is found in the urine. The absence of riboflavin (vitamin $B_2$) blocks these reactions because of failure to replenish FAD. The clinical consequences can be varied, from muscular asthenia to convulsions and to cerebellar symptoms.

If the Krebs cycle is blocked upstream of the succinic acid, there will be a lack of this fundamental substance and the body will try to compensate by synthesizing it from the amino acids leucine and isoleucine. These reactions require the presence of suitable quantities of vitamin $B_{12}$. The amino acid γ-aminobutyric acid (GABA) is also involved in the metabolism of succinic acid. If the level of succinic acid is low, GABA acid is oxidized to succinic acid; if it is sufficient, it is reduced to γ-hydroxybutyric acid.

Succinic acid plays an important role, via its metabolic transformation to mevalonate, as a stimulant of insulin secretion in the pancreatic islets[60] (**Fig. 4.22**).

Some of the reactions that lead to the synthesis of heme (**Fig. 4.23**) occur inside the mitochondrion.

[60] Pomytkin IA, Kolesova OE. Key role of succinate dehydrogenase in insulin-induced inactivation of protein tyrosine phosphatases. Bull Exp Biol Med 2002;133(6):568–570

Heme is then transported to the cytoplasm, where it is conjugated with the polypeptide chain. The first stage of this process involves the condensation (catalyzed by D-aminolevulinic acid synthetase) of glycine with succinyl-coenzyme A, which leads to the formation of D-aminolevulinic acid (which then leaves the mitochondrion). Then, two molecules of D-aminolevulinic acid are condensed, by the action of D-aminolevulinic acid dehydratase, to form porphobilinogen. Four molecules of porphobilinogen are then condensed to form a linear tetrapyrrole, through porphobilinogen deaminase. The tetrapyrrole cyclizes, forming uroporphyrinogen III, which is then transformed into coproporphyrinogen III, by uroporphyrinogen III decarboxylase, which returns to the mitochondrion. Subsequently, via coproporphyrinogen III oxidase, protoporphyrinogen IX is synthesized, which is transformed by protoporphyrinogen IX oxidase into protoporphyrin IX, to which iron is added by ferrochelatase to form the heme group. Thus, the correct metabolism of succinic acid is strictly correlated with good hematopoietic function in the bone marrow.

#### *Homotoxicological Use*

The key role in the production of energy at the oxidative phosphorylation level, the stimulation of insulin secretion, and the relationship with the biosynthesis of heme make succinic acid a metabolite of great interest for cellular anabolic and reparative processes.

There is considerable clinical experience of its use, particularly in convalescent and exhausted subjects in whom the trophic (insulin), antianemic (heme), and energetic (ATP) stimulus is of fundamental importance. According to the individual case, succinic-Injeel acid will be associated with Musculus suis-Injeel, Medulla ossis suis-Injeel, or Cor suis-Injeel. The use of this catalyst in homotoxicological form in patients undergoing chemotherapy has proved very effective, above all for its medullary effects.

#### *General Considerations*

Homotoxicologically, the various homeopathic dilutions of succinic acid act as a specific remedy for the consequences of stress on the body. Succinic acid acts by activating immunological reactivity and the correct production of the hemato-

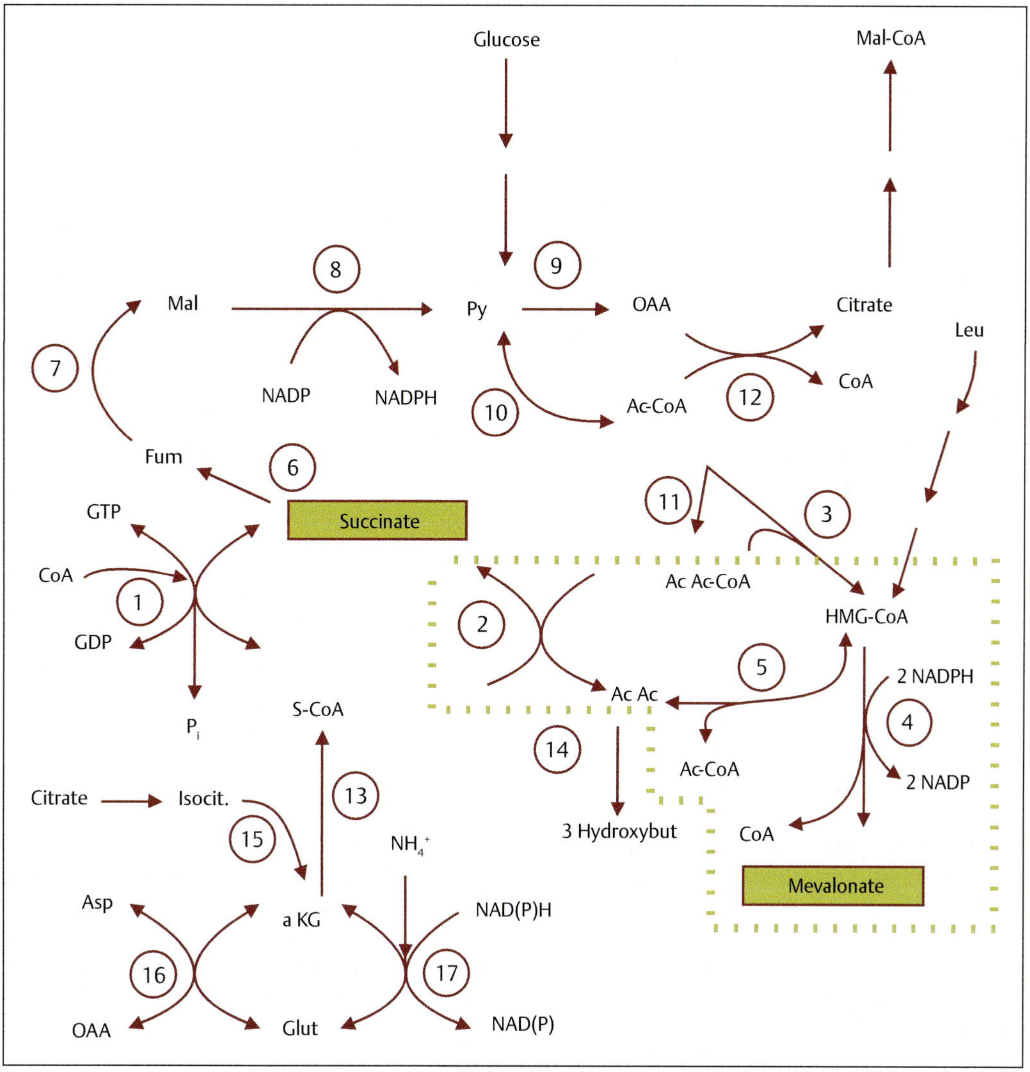


**Fig. 4.22** Succinate/mevalonate transformation.

poietic medullary cells in particular. It is effective in a wide range of indications and is a true and appropriate tissue reactivator[61].

## Baryum Oxalsuccinicum

Baryum oxalsuccinicum is not a true physiological intermediate catalyst; rather, it is a salt considered by Reckeweg as a useful stimulating substance in mitochondrial function, to be used above all in patients in the aging phase. Barium, a typical element of immunological and metabolic activation, well known in research in clinical homeopathy, is associated with succinic acid, a key compound for both the citric acid cycle and oxidative phosphorylation. Calcium and magnesium are important cat-

[61] Bianchi I. Das Krebsproblem im Licht der neuen Erkenntnisse der homöopathischen Wissenschaft und der homotoxikologischen Therapie, *Biologische Medizin*. Baden-Baden: Aurelia Verlag; 1986:20–38

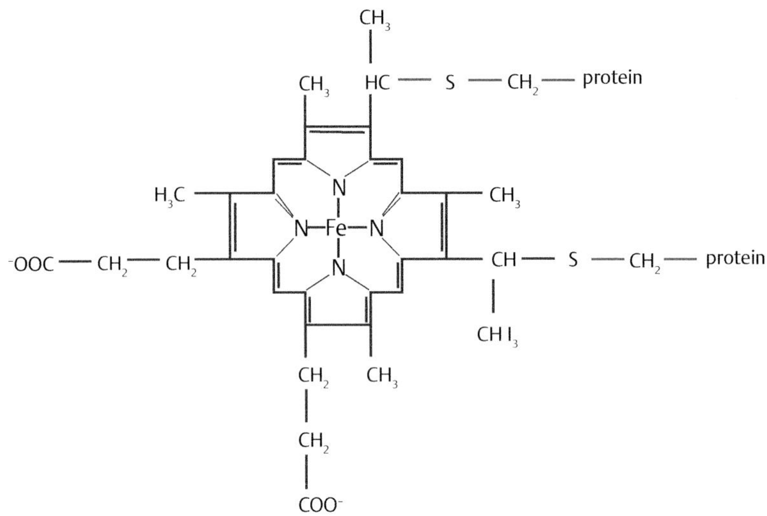


**Fig. 4.23** Heme.

alytic minerals for several mitochondrial enzyme reactions; barium, which is in the same group in the Periodic Table, has a similar activating function. However, this activating function only develops if it is used in small doses.

Similarly to all the other compounds of barium in dilution (Baryta carbonica, Baryta muriatica, and Baryta silicata), Baryum oxalsuccinicum is indicated when the aim is to activate a reaction or stimulate a function. In homotoxicological form, Baryum oxalsuccinicum is typically indicated in all clinical conditions that are related to aging. These conditions include involutional depression, arterial hypertension, gonadal deficit in menopause or andropause, tissue sclerosis, and osteoporosis.

The preparation will be usefully associated with specific organotherapeutic agents (e.g., Funiculus umbelicalis suis and Embryo totalis suis) and with unitary homeopathic agents typical of aging (e.g., Lycopodium, Conium, Sepia, Plumbum, and Selenium). Baryum oxalsuccinicum is, then, a potent activator of antidegenerative preparations with specific or general tissue action.

#### *General Considerations*

Homotoxicologically, the various homeopathic dilutions of barium oxalosuccinate act as a specific remedy for several metabolic and endocrine disorders. The early correction of these changes, which often occur during puberty (particularly during the period of cessation of gonadal function), prevents the large number of pathological conditions related to these instances of hormonal imbalance. The remedy should be used systematically during adolescence and during andropause and menopause[62].

### Acidum Fumaricum

#### *Elements of Biochemistry and Physiopathology*

Fumaric acid (**Fig. 4.24**), in addition to being an important Krebs cycle catalyst, is also a collateral product in the urea–arginine cycle and purine biosynthesis. Pharmacologically, it is a very promising substance. It has been used as an acidulant, an antidermatitic agent, an antihepatocarcinogenic agent, an antioxidant, an antipsoriatic agent, and

---

[62] Bianchi I. *Patologie Geriatriche e Terapia Omotossicologica.* Milan: Guna Editore; 2006

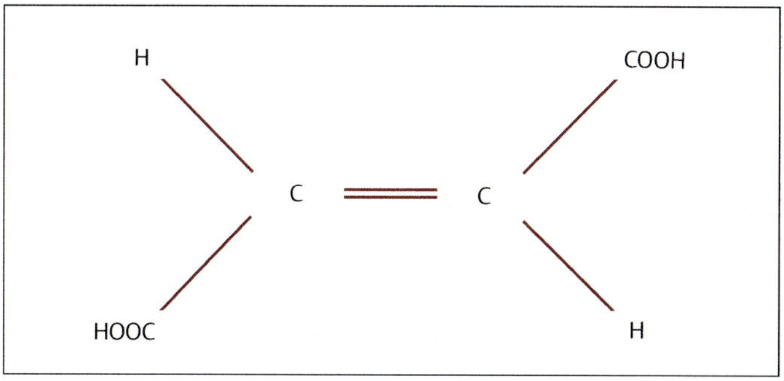


**Fig. 4.24** Fumaric acid (*trans*-but-2-ene dioic acid).

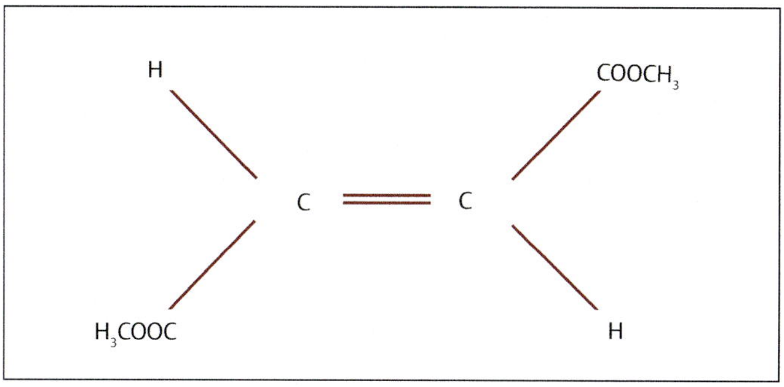


**Fig. 4.25** Methyl fumarate.

an antitumoral agent. In healthy individuals, fumaric acid is produced in the skin after exposure to UV radiation from the sun. Psoriatic patients have a metabolic defect whereby they produce fumaric acid with great difficulty or need prolonged exposure to the sun, and psoralen– UV-A treatment if young.

An important contributory cause of psoriasis is, therefore, the defective synthesis of fumaric acid and the consequent accumulation of intermediate metabolites that seem to be the primary cause of skin damage. However, fumaric acid is a metabolically very active substance and must, therefore, be administered with great caution so as not to cause dangerous blood glucose oscillations, particularly in the hypo direction. In patients with psoriasis, good results have been obtained both with systemic and topical administration of fumaric acid. The methyl ester of the metabolite (**Fig. 4.25**) has been shown to be particularly active. The administration of fumaric acid esters to psoriatic patients has contributed to improvements in about 70% of patients. However, adverse effects (notably hot flashes and gastroenteric disorders) have also been noted in about 60% of patients. These effects are dose dependent and are attenuated during the course of therapy.

Fumaric acid and its esters seem to have a net immunological effect, correcting the disequilibrium toward $T_H1$ of the $T_H1/T_H2$ balance that inhibits proliferation of keratocytes[63].

Animal studies have confirmed a net effect of inhibition of the hepatocarcinogenic stimuli induced by introducing thiocetamide into the diet when fumaric acid (still in small quantities) is also a supplement in the diet.

Despite claims of great efficacy, therapeutic doses of conventional drug based on fumaric acid esters often have serious side-effects at the hepatic, renal, and medullary levels.

---

63 Nieboer et al. Systemic therapy with fumaric acid derivates: new possibilities in the treatment of psoriasis. *J Am Acad Dermatol.* 1989Apr;20(4):601–608

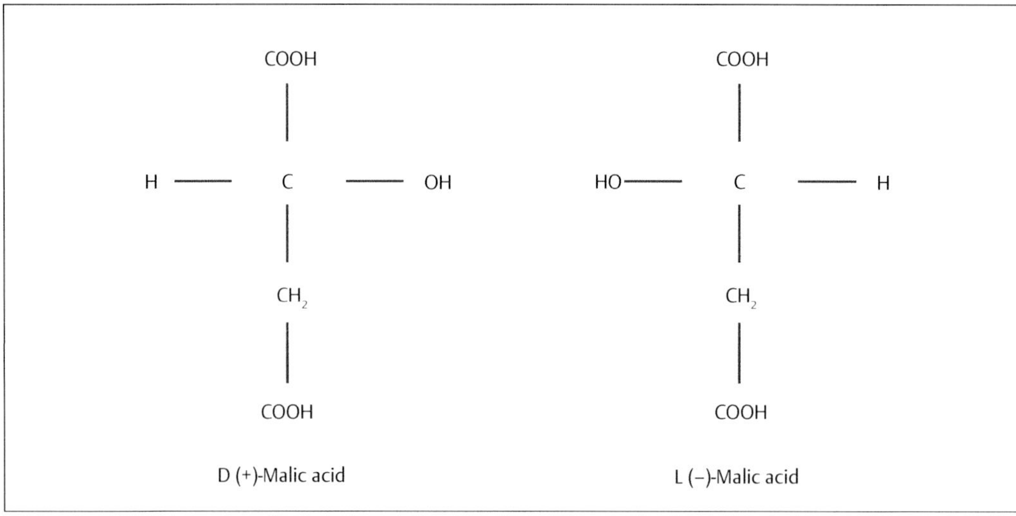


**Fig. 4.26** D-Malic acid and L-malic acid.

### *Homotoxicological Use*

Fumaric acid in homotoxicological form has been of great interest in several clinical studies. Usually when small doses of a substance are administered, the indication is not so wide and, therefore, Acidum fumaricum-Injeel is not a drug for all forms of psoriasis. The regulatory effect of the Injeel preparation on the metabolism of fumaric acid is such that several biosynthetic pathways are activated, such as those of thyroxine and arginine synthesis, which induce both hepatic and renal metabolic activation. The synthesis of thyroxine and, consequently, of phenylalanine implies that it is this cutaneous photosensitization that is responsible for the improvement of the psoriatic subject. However, it is one thing to force the metabolism of fumaric acid with megadoses, causing toxic accumulations at several levels, and another for a biochemical reaction to be activated with a suitable metabolic stimulus. Acidum fumaricum-Injeel is an optimal therapeutic alternative, with no side-effects, to dangerous massive allopathic doses. This therapy will be particularly effective when psoriasis is concomitant with a dysmetabolic block and is therefore associated with hypercholesterolemia and poor hepatic function.

These biochemical considerations also explain the efficacy of this homotoxicological catalyst in various ailments characterized by the accumulation of toxins, such as dyslipidemia, hypertension, obesity, and involutional depression. The consequences of food intolerance can also be buffered by homotoxicological activation of the metabolism of fumaric acid, because the basis of these dysfunctions is often the inability to metabolize some substrates within the Krebs cycle and the urea cycle, both conditioned by fumaric acid.

### *General Considerations*

Homotoxicologically, the various homeopathic dilutions of fumaric acid act as a formidable therapeutic instrument for activating cleavage of toxins and metabolism of fats at all levels. Fumaric acid is indicated in every situation of impregnation, tendency to chronicization, sclerosis, and senile degeneration, generalized or organ specific. It can be a useful tool for the prevention of all degenerative diseases[64].

64 Bianchi I. Il biocostituzionalismo omotossicologico: suoi influssi sul metabolismo. *La Medicina Biologica*. 2006;(4):3–12

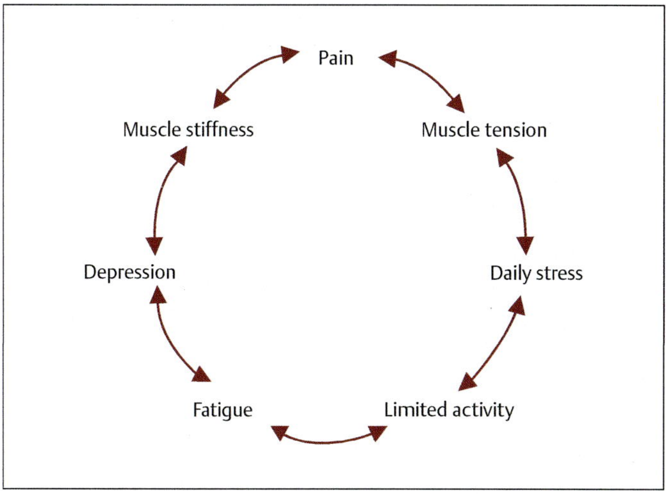


**Fig. 4.27** The fibromyalgia cycle.

## Malic Acid

### *Elements of Biochemistry and Physiopathology*

Malic acid (**Fig. 4.26**) is a natural constituent of many fruits and vegetables and is degraded when fermentation occurs. In the body, it derives both from diet and from synthesis in the Krebs cycle. In aerobic conditions, malic acid is transformed into oxaloacetic acid, which again triggers the Krebs cycle. In anaerobic conditions, malic acid is metabolized to both succinic acid and oxaloacetic acid, consuming the damaging cellular accumulation of reducing factors and allowing glycolysis (which is blocked by these substances) to be restarted. In some pathological conditions, such as fibromyalgia, it is the tissue hypoxia that blocks the production of cell energy and induces tissue damage.

In animal experiments, it was shown that with intense physical exercise malate is consumed almost exclusively in the tissue (i.e., the glucose is metabolized directly to malic acid, which, in turn, is oxidized to produce energy). Therefore, there is reason in its ponderal integration in athletes who want to maximize their physical performance. The lack of malic acid in the tissue (i.e., the deficit of conversion of glucose into malic acid) seems to be one of the principal causes of physical fatigue on exertion. Supplementation, still in the experimental animal, of small amounts of malic acid allows an increase in mitochondrial malate, cellular respiration, and the production of ATP. In anaerobic conditions, the requirement and consumption of malic acid increases; the malic acid is supplied by the activation of gluconeogenesis and muscle protein catabolism, which will induce a reduction in muscle mass and damage. In another study on the oral administration of malate in the rat, an improvement was noted in resistance to anaerobic stress with no increase in consumption of carbohydrates and oxygen.

Malic acid is the only Krebs cycle metabolite positively correlated with physical activity, which induces cellular accumulation of malic acid. In humans, constant physical exercise induces a significant increase in the enzymes involved in malic acid metabolism. Because of the evidence of the role played by malic acid in the production of energy, particularly during hypoxia, the effects of the ponderal supplementation of malic acid in patients suffering from fibromyalgia have been investigated (**Fig. 4.27**). A subjective improvement in pain is already noticeable 48 hours after the administration of 1200 to 2400 mg of malic acid. Magnesium was also associated with these patients in view of the typical absence of this element in those with this syndrome; however, the effects on pain were clearly attributable to malic acid (i.e., when malic acid administration was suspended, the pain returned) (see **Fig. 4.28**)[65].

---

[65] Abraham GE, Flechas JD. Malic acid reduces muscle pain while increasing endurance in CFS and FM. *J Nutr Med.* 1992;3:49–59

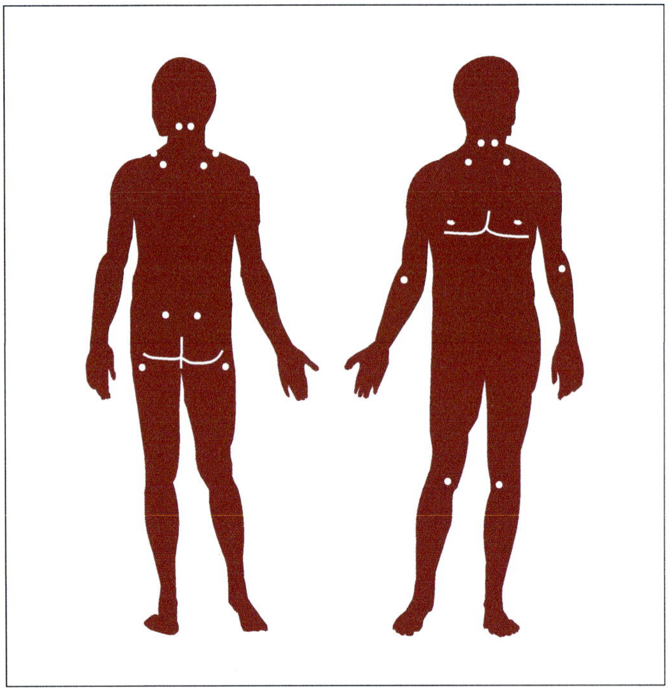

**Fig. 4.28** Typical painful points in fibromyalgia.

In fact, any condition that is hypoxic and, therefore, linked to circulatory or respiratory problems can benefit from the administration of malic acid (e.g., chronic fatigue syndrome, diabetes, and atherosclerosis).

### *Homotoxicological Use*

Acidum DL-malicum in homotoxicological form is above all of great interest in chronic degenerative pathological conditions and in particular in elderly and hyporeactive patients. Senile diabetes, with its complex metabolic changes that are fundamentally influenced by the malic acid molecule, is one of the key indications of the homotoxicological phial. In addition, all conditions in which it should be reactivated are specific indications for the use of Acidum DL-malicum-Injeel. Sometimes, reactivation can be brusque and revival of so many linked metabolic pathways may give the patient symptoms such as fever, hot flashes, or irritability. Therefore, acidum DL-malicum-Injeel, particularly in the elderly, is to be used cautiously; it is better if use is associated with fumaric acid, for a more balanced and harmonious stimulus of the intermediate metabolism.

Activating the metabolism of malic acid often means resolving hypoxic and local situations. These are, therefore, an indication for homotoxicological injection at points of fibromyalgic pain or of hypoxic tissue pain due to any other reason, from diabetes to atherosclerosis or anemia. Clinical results are often speedy and surprising, particularly if the preparation is combined with unitary and/or organotherapeutic components appropriately selected on the basis of the individual clinical case.

### *General considerations*

Homotoxicologically, the various homeopathic dilutions of Acidum DL-malicum act as a remedy that must be used systematically for all patients who say they are well but who, in fact, are to be considered at risk from some subclinical chronic degenerative disease.

Malic acid specifically activates the transport of electrons at the cellular level, reclaiming oxygen in relatively hypoxic tissues. In the case of the advanced impregnation or neoplastic phase, phases in which the body physiologically reduces the oxygen supply to halt cell growth, general

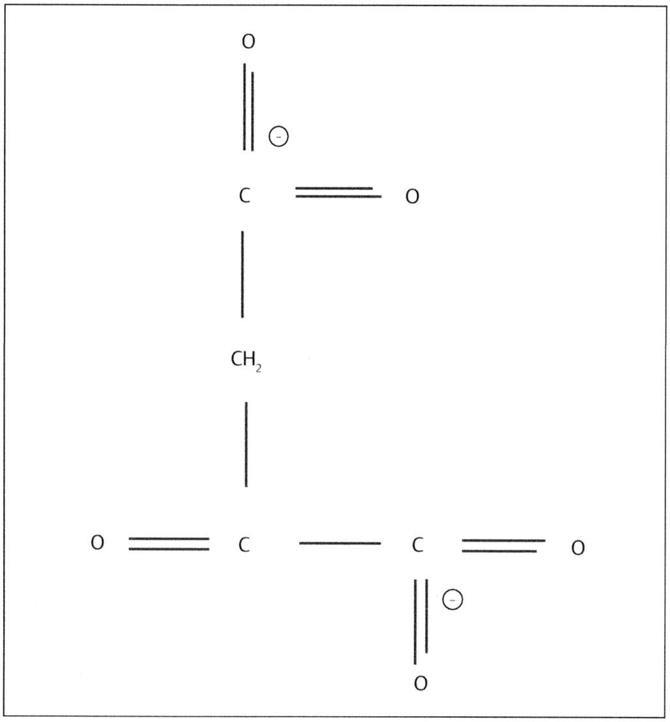

**Fig. 4.29** The oxaloacetate anion.

damage can occur on activation of the metabolism. The tissue toxins tend, in fact, to absorb a large amount of oxygen by accelerating cell proliferation. It is, however, very useful for the protection of healthy tissues; therefore, repeated administration must be performed, always in association with fumaric acid. The metabolism of malic acid activates the detoxifying and oxygenating function of the muscles. Citric, malic, and succinic acids are necessary for the drainage of toxins produced by muscular work. Many forms of rheumatism are connected with a metabolic block of one of these acids.

When the injection of malic acid is followed by fever, this is an index of change in carbohydrate metabolism (e.g., prediabetes, latent diabetes, or a chronic change in the pancreas). Malic acid is an important drug for tissue dyscrasia, for impregnation, and for chronic inflammation. It has a particularly favorable effect on the myocardium[66].

[66] Bianchi I. Il ruolo dell'omotossicologia nella prevenzione antisenile. *La Medicina Biologica.* 2004;(4):3–9

## Natrium Oxalaceticum

### *Elements of Biochemistry and Physiopathology*

Oxaloacetic acid (**Fig. 4.29**) can derive from the catabolism of sugars via pyruvate, by the catabolism of amino acids and malate. In diabetic subjects, in whom there is an inability to use sugars, or in those who are fasting, oxaloacetate used for gluconeogenesis is available in fairly meager amounts.

The transformation of malate into oxaloacetate is inactivated in the presence of meager amounts of glucose. Consequently, the Krebs cycle is not activated and the unused acetyl-coenzyme A is used for the catabolism of fats and largely for the formation of ketone bodies. The overload of ketones induces a state of metabolic acidosis that creates various tissue problems. In summary, when we are faced with a meager availability of glucose and large amounts of fatty acids, an excessive amount of ketone bodies is formed at the hepatic level. Low levels of pyruvate also reduce the availability of oxaloacetic acid and overload the Krebs cycle with acetyl-coenzyme A derived

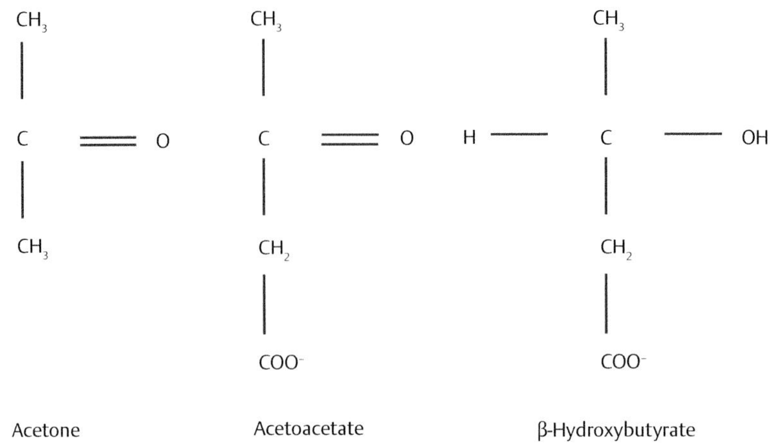


**Fig. 4.30** Acetone, acetoacetate, and β-hydroxybutyrate.

from the degradation of the pyruvic acid and above all from the degradation of the fatty acids. These molecules combine to form acetoacetate, which, in turn, is converted into β-hydroxybutyrate or acetone (**Figs. 4.30, 4.31**). The liver is incapable of metabolizing the ketone bodies, which are, therefore, transported to the muscles and the brain. The excess is expelled from the kidney as the condition of ketonuria, an index of metabolic acidosis.

The acetone cannot be used and is expelled by the lungs (see **Fig. 4.31**). The characteristic acetone odor of a subject (halitosis that smells of fruit) also indicates a state of metabolic acidosis. From the knowledge obtained via biochemistry, some practical therapeutic considerations emerge: (1) Combining the administration of pyruvate and oxaloacetate is useful. (2) Oxaloacetate therapy should be integrated in various situations, from common and benign ketonemia in infancy to the much more serious ketonemia found in diabetes and renal insufficiency.

The heart and the nerve cells use glucose and ketone bodies as fuel, transforming them into nicotinamide–adenine dinucleotide (NAD) and acetoacetate. The latter can be activated by succinyl-coenzyme A to acetoacetyl-coenzyme A and succinate. The acetoacetyl-coenzyme A reacts with thiolase to form two units of acetyl-coenzyme A, which are metabolized in the Krebs cycle.

In the final analysis, it is clear that the balanced availability of oxaloacetate brings about the proper functioning of the Krebs cycle and, therefore, the correct use of acetyl-coenzyme A, antagonizing the accumulation of fat, and the excessive production of ketone bodies, inducing metabolic acidosis. These considerations also allow us to understand what should be an appropriate diet both to limit accumulation of fats and to avoid metabolic dysfunction[67] (**Fig. 4.32**).

We can see how a diet rich in carbohydrates (e.g., the Mediterranean diet) should always accurately take account of the oxaloacetate levels and how fasting is very dangerous metabolically, particularly in the case of an absolute or relative absence of oxaloacetate. Speaking biochemically, as has also been widely demonstrated in clinical practice, the ideal diet for reducing the accumulation of fats is one rich in proteins and poor in fats and sugars. Surely this is the only dietetic regimen that offers a valid mechanism for the reduction of fat mass. Amino acids in the diet supply oxaloacetate and are used for gluconeogenesis; the fat is

[67] Lieberman M, Marks AD. *Basic Medical Biochemistry.* Baltimore: Lippincott Williams & Wilkins; 2008

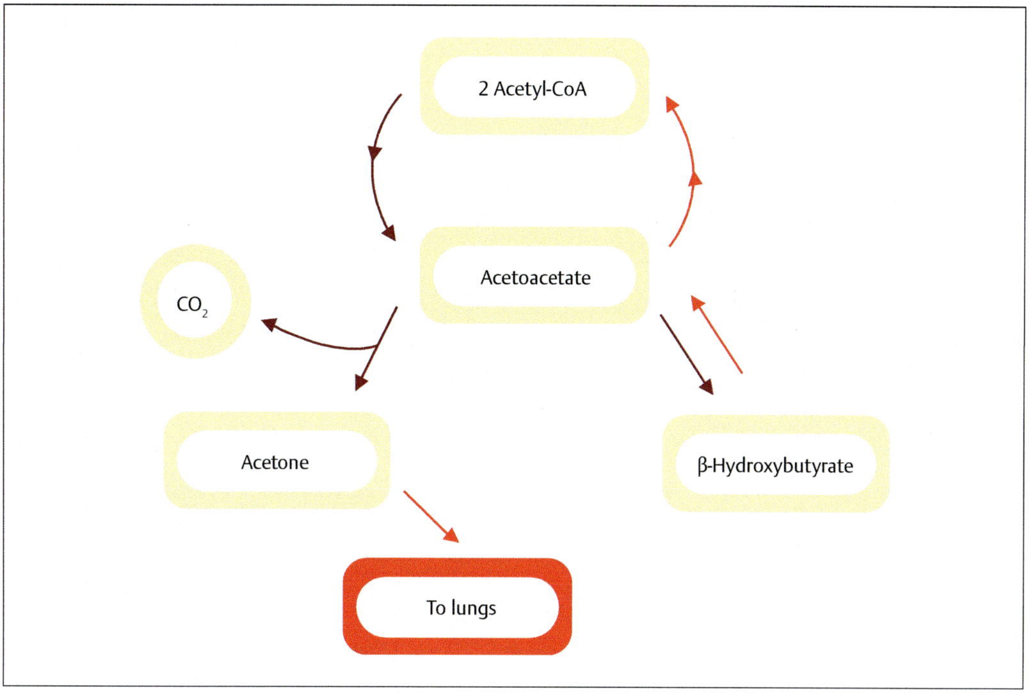


**Fig. 4.31** The formation of acetone and hydroxybutyrate from excess acetyl-coenzyme A, and the expulsion of acetone via the lungs.

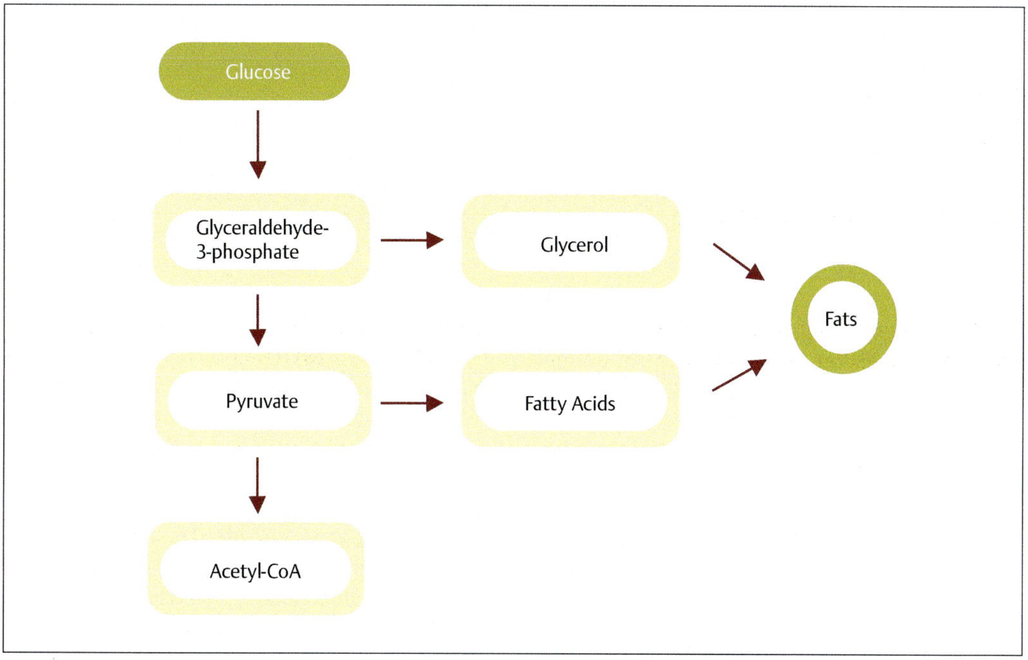


**Fig. 4.32** Carbohydrate-rich diet in the absence of oxaloacetate causes the accumulation of fats.

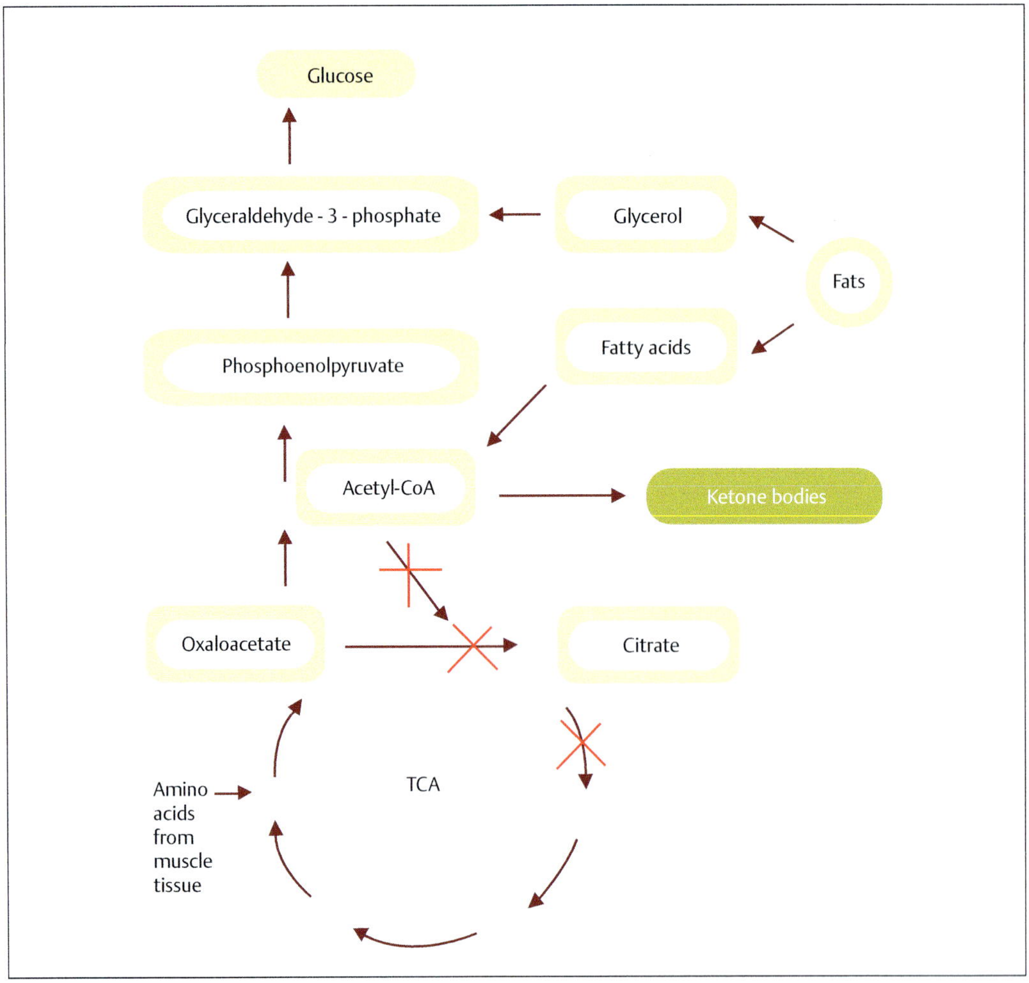


**Fig. 4.33** Result of oxaloacetate deficiency.

used to produce acetyl-coenzyme A and, therefore, subsequently ATP in the Krebs cycle. An example of a good application of this diet might be a regimen based on a wide consumption of fish. However, this diet is not devoid of problems; in particular, there exists the possible accumulation of mercury and the impact of an excessive contribution of omega-3 lipids on coagulation. However, the best way to reduce fat mass is still a low-fat diet combined with physical exercise (see **Figs. 4.33, 4.34, 4.35**).

It is also important to remember the considerable role of oxaloacetate in the urea cycle, the primary biochemical sequence brilliantly discovered by Krebs (**Fig. 4.36**). Oxaloacetate has a catabolic role (i.e., it is consumed in the citric acid cycle in order to produce energy in the form of ATP) and a considerable anabolic role (i.e., it is an important building block in the synthesis of various amino acids that have a fundamental role in trophism and in muscle mass growth). The amino acid arginine is formed from oxaloacetate (passing through aspartate). Aside from its fundamental anabolic role in tissue growth, arginine has a role at the catabolic stage because, on entering the urea production cycle, it leads to the production of nitric

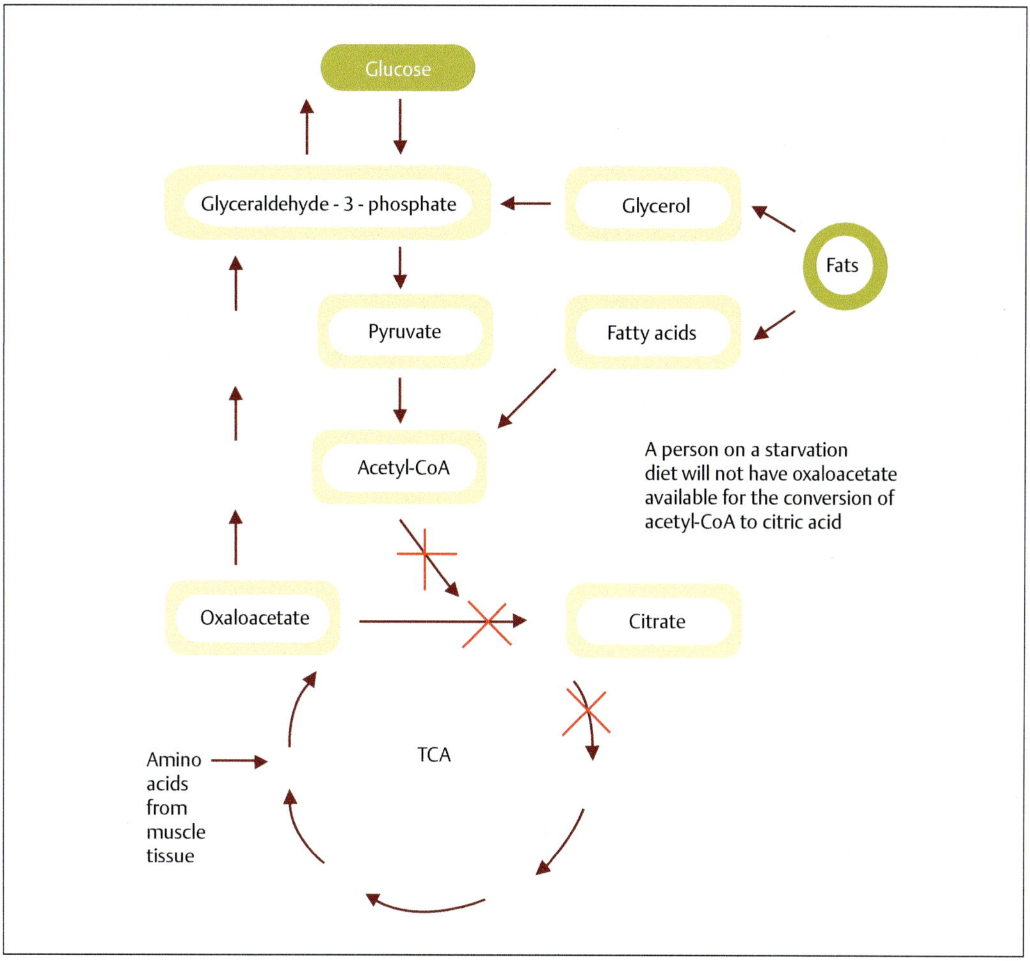


**Fig. 4.34** Fasting: in the absence of oxaloacetate, fats cannot be metabolized in the Krebs cycle. Oxaloacetate will be derived from the degradation of amino acids with consequent muscular atrophy.

oxide gas, to which greater and greater importance for cardiovascular function has been attributed in recent years.

### *Homotoxicological Use*

The sodium salt of oxaloacetic acid in homotoxicological form is a therapeutic instrument of fundamental importance. Oxaloacetate itself is the critical factor that can nullify the functioning of the Krebs cycle function in the body, as can occur in contemporary humans who, because of sedentary habits and a fat-rich diet, produce excessive and unbalanced amounts of acetyl-coenzyme A. The body has feedback mechanisms to stimulate the availability of oxaloacetate, particularly when the level of acetyl-coenzyme A increases. However, these mechanisms do not always function for the best. Thus, the stimulation of the Krebs cycle reactions with the various catalysts in homotoxicological form can be a key therapy for metabolic rebalancing. In particular, the absence of oxaloacetate in the liver, the organ of production and storage of energy, leads to lack of glycolysis, excessive oxidation of fatty acids, exaggerated and unbalanced production of acetyl-coenzyme A, and preferential use of the little oxaloacetate for amino acid synthesis.

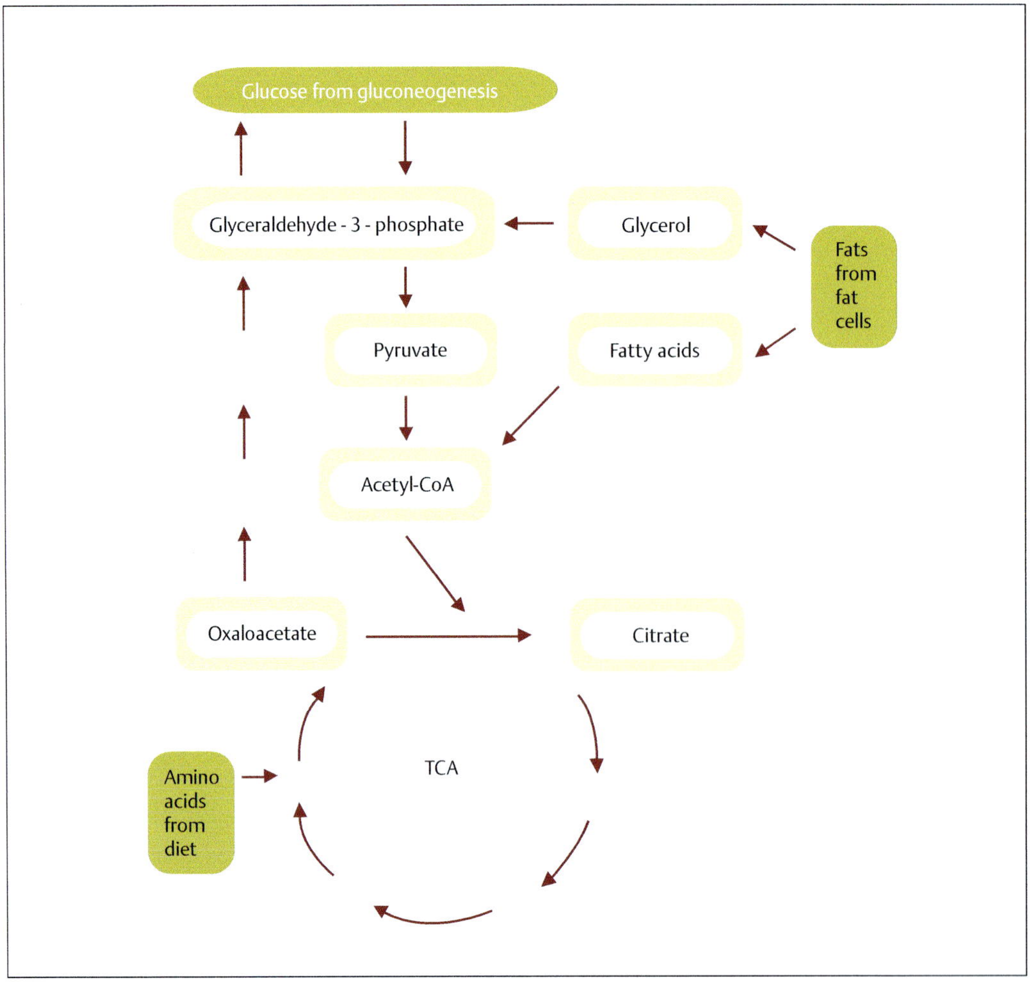


**Fig. 4.35** A deficiency of oxaloacetate always induces a serious risk of metabolic acidosis.

Therefore, in the final analysis, the absence of oxaloacetate leads to poor production of energy by the cell, which opens the way to a whole series of pathological conditions. These conditions are mostly, but not only, of the immunological type, given the essential good availability of ATP in the cell for the synthesis of interferons and cytokines, which activate and regulate the function and proliferation of the cells of the immune system. The clinical situations in which the administration of Natrium oxalaceticum in homotoxicological form is useful for stimulating and rebalancing this metabolism include the following:

- Accumulation of endogenous and exogenous toxins
- Aftereffects of viral and bacterial diseases
- Tendency to chronicization of acute pathological conditions
- Excessive tendency to contract infectious diseases and common cold
- Food allergies and intolerances
- Psychophysical prostration
- Deficit of sexual function and erection in particular
- Immunological deficit with compensatory lymphatic hypertrophy
- Chronic catarrh

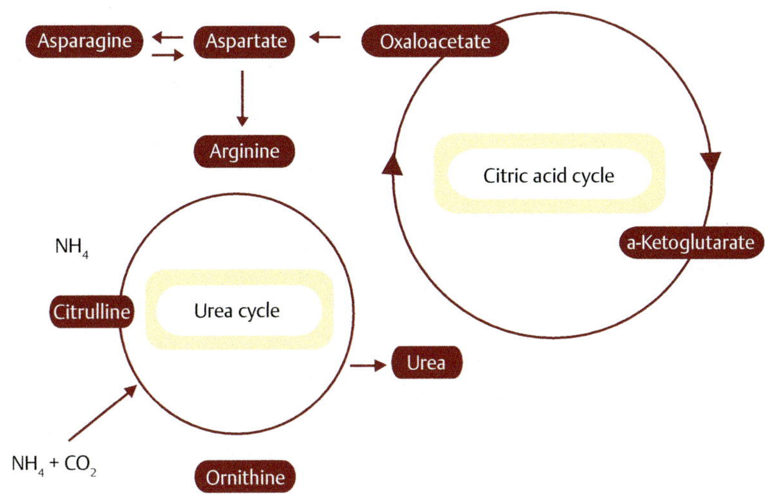


**Fig. 4.36** The role of oxaloacetate in the urea cycle.

*General Considerations*

Natrium oxalaceticum is a key catalyst in triggering the Krebs cycle. It is useful in all cases in which there is an underlying cell energy deficiency and, consequently, an immunological deficit with a tendency toward unhealthy recurrence and chronic inflammatory pathological conditions. Its fundamental role in the citric acid cycle makes it useful at any age, both for normal problems of childhood (due to feeding disorders, even temporary) and for more serious problems (such as tissue acidosis in adults). Tissue acidosis is linked to metabolic imbalances, such as obesity and diabetes; through immunological inhibition, these imbalances can lead to fairly serious and unfortunately common chronic and degenerative pathological conditions, such as neoplasias.

*Practical Application of the Homotoxicological Therapy of Changes in the Krebs Cycle*

It clearly emerges that the Krebs cycle, in its various stages, must preferably be stimulated by the use of microdoses of substance, the only ones to guarantee a nonsuppressive stimulus. However, although the microdose of a substance is important it is only one stimulatory factor. The fundamental role of combined Injeel dilutions is noted; these dilutions stimulate and brake to the same extent and, therefore, rebalance the function. In particular, in patients today, we are seeing not so much problems of a deficit type but of functional blockage by toxin or other overload. Here then is the fundamental therapeutic role of combined Injeel dilutions. Therefore, the correct approach of mitochondrial therapy in a patient in the chronic phase will be the use of the Injeel forms of the individual catalysts initially and, possibly, administration in sequence of the collection of the catalysts themselves. After rebalancing of the citric acid cycle, it can at various times be stimulated with periodic administrations of small combined doses of intermediate catalysts of the Krebs cycle, such as those contained in the Coenzyme compositum preparation. This explains why Coenzyme

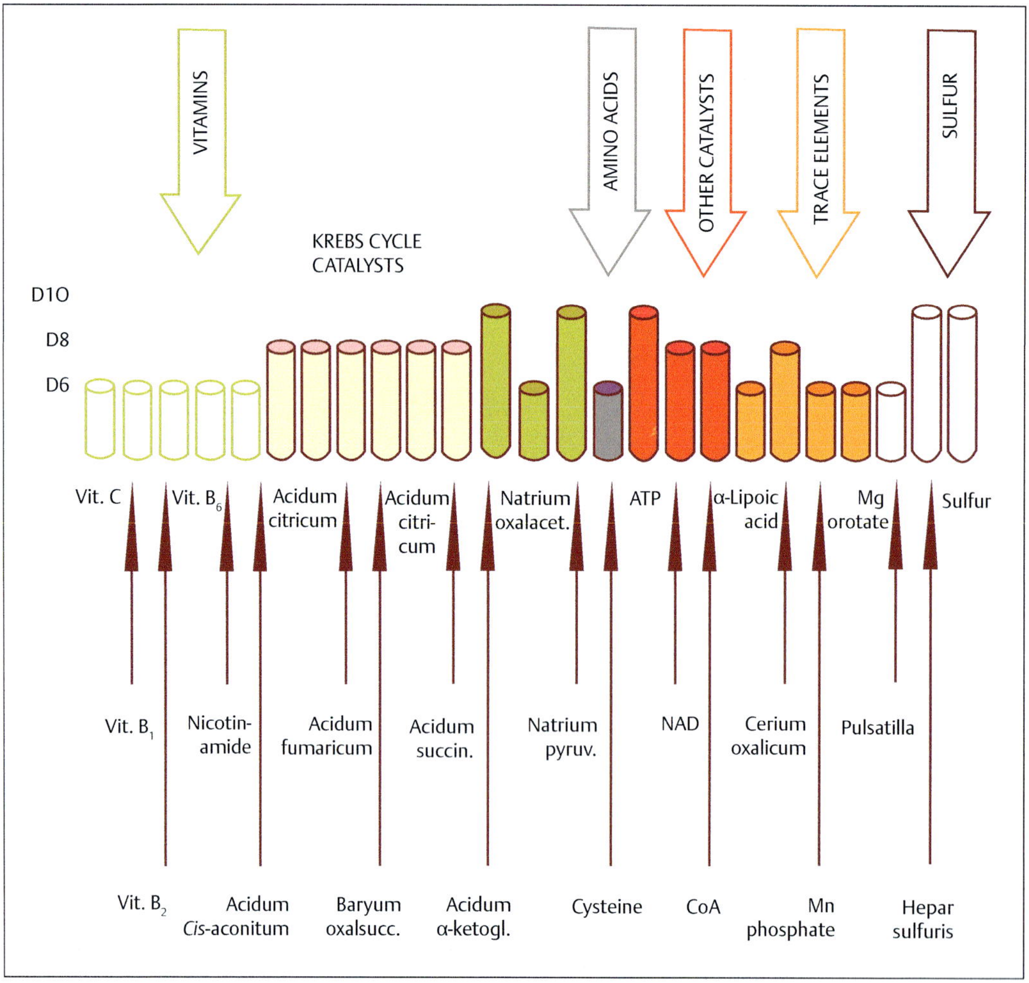


**Fig. 4.37** Diagram of the practical use of the Krebs cycle intermediate catalysts.

compositum is often not so effective for elderly patients and why, on the other hand, the same drug has had considerable efficacy in children, in whom it can be useful in some cases to stimulate the cycle or the production of energy in the absence of blockages and metabolic obstructions[68]. See **Fig. 4.37**.

[68] Bianchi I. *Homeopathic – Homotoxicological Repertory Homotoxicological Materia Medica*. Milan: Natali & C. s.a.s.; 1993

*Patients' Global Clinical Classification*

- **Patient in humoral phase:** Coenzyme compositum
- **Patient in cellular phase:** Single package of Krebs cycle catalysts; Coenzyme compositum

*Diagnosis of Specific Pathology*

- Unitary catalyst or catalyst pair
- Specific trace elements
- Specific quinone
- Specific suis organotherapeutic agents

### *Concluding Summary on Catalysts*

- The tricarboxylic acid cycle is the fundamental pathway of cell energy metabolism.
- The production of energy is not, however, the only function performed by the Krebs cycle, which produces, in addition, several intermediate metabolites necessary to the organic anabolic processes.
- The products of the digestion of carbohydrates, lipids, and proteins merge (after transamination) in the cycle and their progressive degradation, coupled with the respiratory chain, produces chemical energy. Hydrogen, originating from the oxidation processes, is burned along the respiratory chain, in turn producing energy via the enzyme cytochrome oxidase, then combining with molecular oxygen to give water.
- The energy produced is partly thermal and partly chemical, stored in the form of ATP.
- The Krebs cycle is located, at the cellular level, in the mitochondria, even though some enzymes that compose it are cytoplasmic. The mitochondrial location is fundamental for coupling with the respiratory chain, into which hydrogen produced by the Krebs cycle flows. A block in any site of this complex system causes all reactions to stop.
- In sequential order, the Krebs cycle enzymes are as follows: citrate synthetase, aconitase, isocitrate dehydrogenase, α-ketoglutarate dehydrogenase, succinyl coenzyme A dehydrogenase, succinate dehydrogenase, and malate fumarase dehydrogenase.
- Therapy with the homeopathized Krebs cycle factors has the power to send an impulse stimulating and/or regulating the intermediate functions of this fundamental metabolic chain.
- The role that the Krebs cycle (or the tricarboxylic acid cycle) represents as the metabolic ring through which all dietary substrates (proteins, carbohydrates, and lipids) enter energy metabolism should be emphasized.
- The function of the Krebs cycle is that of degrading molecules to obtain $CO_2$ and, above all, four pairs of hydrogen atoms (the so-called reducing equivalents) that then enter the electron transport chain and arrive at the terminal, where they will find the molecular oxygen that they will reduce to $H_2O$ (water) *with the production of energy in the form of ATP.*
- The mechanism of action of the Krebs cycle intermediates in homeopathic form is attributable to the laws of enzyme kinetics: every intermediate of the cycle represents the substrate for the specific enzyme that has to catalyze its transformation into the next intermediate (e.g., citric acid represents the substrate for catalyzed reaction of the enzyme aconitase, the role of which is to transform the citric acid into *cis*-aconitic acid). The homeopathic form (i.e., extremely diluted) itself functions as an inducer, accelerating the speed of reaction. (Cf. the Michaelis–Menten law: in a catalyzed enzyme reaction, the speed of reaction is inversely proportional to the concentration of the substrate.)
- The Krebs cycle intermediates are not only able to abolish the blockages (e.g., pyruvate accumulation that would slide toward the anaerobic pathway, with marked slowing of final oxidation) but to actually restore the operation of the Krebs cycle, *by the induction of the corresponding enzymes.*

## *Oxidative Phosphorylation and the Quinones*

**Introductory Ideas on Oxidative Phosphorylation**
Oxidative phosphorylation is the process that occurs in the inner membrane of the mitochondrion, by which redox potential energy contained in the various oxidizable substrates catabolized in the Krebs cycle is transformed into phosphate bond energy. In practice, for each pair of electrons that are transported along the whole respiratory chain (i.e., for every molecule that is oxidized in the cycle), some adenosine diphosphate (ADP) and inorganic phosphate (Pi) molecules are transformed into the same number of ATP molecules. These are two paired chemical events, one oxidative and one phosphorylative. Phosphorylation uses 40% of the energy produced by the oxidative chain.

In aerobic conditions, oxidative phosphorylation provides for the formation of most of the cell's ATP. During oxidative phosphorylation, both the reduced form of NAD (NADH) and the reduced form of FAD ($FADH_2$), produced by glycolysis and by the Krebs cycle, are used. Through a multienzyme complex with the functions of a transport chain, electrons are removed from the NADH and $FADH_2$ and, after a series of intermediate transfers, are transferred to molecular oxygen, which is reduced to water. During electron transfer, the various transport proteins undergo conformational changes that allow the transfer of protons from the matrix to the intermembrane space against a concentration gradient.

In the mitochondrion, we can isolate as many as four polyprotein complexes responsible for the transport of the electrons:

1. **Complex I (NADH dehydrogenase):** This complex contains at least 30 different polypeptides, one flavoprotein, and nine iron–sulfur centers. Three or four protons are transferred for each pair of electrons transferred.
2. **Complex II (succinate dehydrogenase):** This complex, in addition to catalyzing a Krebs cycle reaction, allows the transfer of electrons to FAD and to ubiquinone but does not allow the transfer of protons.
3. **Complex III (cytochrome-*c* reductase):** This complex contains about 10 polypeptides and heme groups and one iron–sulfur center; it allows the transfer of electrons from reduced ubiquinone to cytochrome *c* and transfers four protons for each pair of electrons.
4. **Complex IV (cytochrome-*c* oxidase):** This complex contains at least 13 polypeptides; it allows the transfer of electrons from cytochrome *c* to oxygen and the displacement of protons even if the number is not very clear (perhaps four for reduced oxygen).

Subsequently, the protons again pass through the inner membrane in a facilitated diffusion process, via the enzyme ATP synthetase, which in this way obtains sufficient energy to produce an ATP molecule, transferring a phosphate group to an ADP. A pair of electrons removed by NADH is able to release sufficient energy to produce three ATP molecules, whereas two are obtained with one electron pair obtained from $FADH_2$.

It can be deduced from this that, in vitamin supplementation, vitamin $B_3$ is much more active than vitamin $B_2$ in producing energy in the cell; it can also be understood how an excess of this vitamin, as often used in hypercholesterolemia, leads to hepatic damage linked to the consequent hyperproduction of free radicals. We must, therefore, accurately determine the supplementation of vitamins $B_2$ and $B_3$ in relation to the stage of the patient's intoxication; and if we want to force oxidative phosphorylation, we must remember to add antioxidants such as vitamin E, vitamin C, or preferably lipoic acid.

Both glycolysis and oxidative phosphorylation allow as many as 36 molecules of ATP to be obtained for each glucose molecule used (although this value can also vary according to the intracellular ATP/ADP ratio). The importance of the transfer of protons across the inner mitochondrial membrane in the synthesis of ATP, a mechanism defined as *chemiosmotic*, was identified in 1961 by Peter Mitchell, who was awarded the Nobel Prize for chemistry for this finding in 1978. In 1997, the same prize was awarded to Paul D. Boyer and John E. Walker for clarifying the mechanism of action of ATP synthetase. There are cell studies on animals and clinical evidence that demonstrate that whenever the oxidative phosphorylation mechanism is blocked, as occurs in aberrant cells such as cancer cells, it can be rapidly reactivated by administering parabenzoquinone, a compound with a dehydrogenation capacity superior to that of physiological ubiquinone. The study was performed by assessing the use of glucose by the lymphocytes of patients with cancer. Parabenzoquinone seems to play a role in reducing both the side-effects of chemotherapy and the metastatic tendency of the tumor, thanks to the reestablishment of a normal metabolism in the cells[69].

#### Homotoxicological Therapy of Changes in Oxidative Phosphorylation

In homotoxicology, the therapeutic quinone must be chosen according to two fundamental notions:

1. **Identifying the need to stimulate or slow down the process of oxidative phosphorylation.** In the presence of excessive oxidative processes, we will administer quinones such as hydroquinone, which has outstanding reducing ability. In an extremely hyporeactive patient

[69] Alberts B, Johnson A, Lewis J, Raff M., Roberts K, Walter P. *Molecular Biology of the Cell.* 4th ed. New York: Garland Science; 2001

who tends to accumulate toxins on account of defective cellular oxidation, we will administer quinones with a high oxidant power, such as parabenzoquinone, anthraquinone, and naphthoquinone.
2. **Assessing the redox potential of each quinone.** When going from anthraquinone to naphthoquinone to parabenzoquinone, the redox potential increases progressively. The more advanced the pathological phase, the higher the redox potential of the quinone used in homotoxicological doses must be. We double the potential to stimulate protection mechanisms when dealing with free radicals, in which oxidative phosphorylation is accelerated by an abnormal toxic quinone, and to powerfully stimulate the oxidative processes and, consequently, the production of cell energy where there is a blockage of oxidative phosphorylation. In this last case, we will always add antioxidants such as vitamin E.

An interesting method for correctly selecting a quinone is that of classifying the patient from the neurovegetative viewpoint: it is logical to stimulate cell oxidation with a quinone such as naphthoquinone in the patient with a carbonic-phlegmatic condition who otherwise often has inflammatory problems of the lungs, whereas it will be more prudent to administer hydroquinone in a subject with an accelerated phosphoric irascible condition.

Ubiquinone compositum is an extremely interesting global stimulant of the mitochondrial quinone function, an antiaging remedy par excellence. Combining the drug with powerfully oxidant quinones and discrete amounts of reducing quinones makes it balanced and global in its action. Because Reckeweg had intuited the specific interrelation between the Krebs cycle and oxidative phosphorylation, he also inserted into ubiquinone compositum Natrium oxalaceticum and coenzyme A factors, triggering the Krebs cycle.

However, excessive quinone stimulation can induce a surplus of free radicals that could cause cell genomic damage. Therefore, the drug includes (in small amounts) a whole series of phytotherapeutic agents that have a protective role towards the DNA structure, including colchicum, podophyllum, and conium.

### Characteristics of Quinones

Quinones are biochemical substances characterized by aromatic rings with 2 ketone substitutions. The name was coined in 1838 by Justus von Liebig, who synthesized them by the oxidation of quinic acid during studies on cinchona bark. They are derived from the oxidation of diphenols; this reaction is reversible by the action of reducing agents. They are ubiquitous in nature and are characteristically very reactive substances because of their strongly unsaturated nature. They are colored substances responsible for the oxidative browning that we see on cut or bruised fruit and vegetables and are found in human skin as intermediates in the synthesis of melanin from the amino acid tyrosine.

Conversion from a quinone to its reduced form (**Fig. 4.38**) is a fairly easy reaction precisely because of the considerable reactivity of these substances. Knowledge of the specific redox potential of each pair of quinones is very important for understanding their role in biology (**Fig. 4.39**).

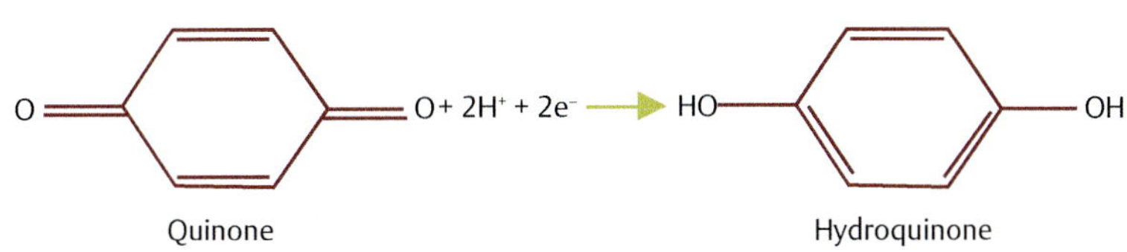


**Fig. 4.38** Redox reaction of quinone.

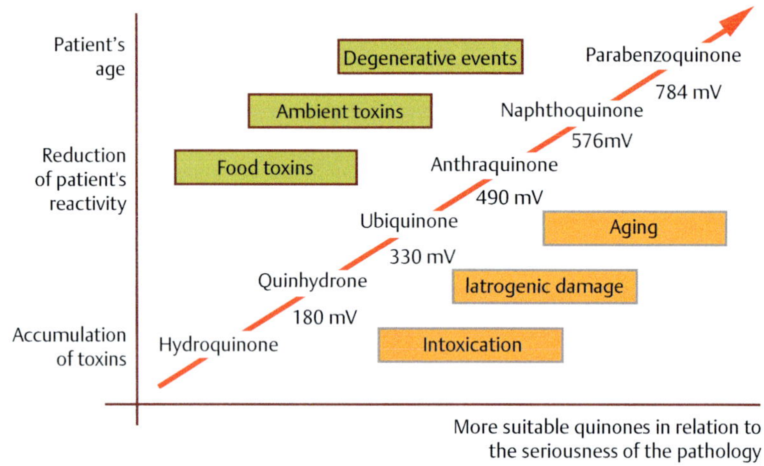


**Fig. 4.39** Redox potentials of some quinones.

*Redox potentials*

| | |
|---|---|
| $NADH \leftrightharpoons NAD^+ + H^+ + 2e^-$ | $-0.32\,V\ (-320\,mV)$ |
| $H_2O \leftrightharpoons \frac{1}{2}O_2 + 2H^+ + 2e^-$ | $+0.82\,V\ (+820\,mV)$ |

On the one hand, there is physiological ubiquinone (coenzyme Q), which is specific for the correct activation of mitochondrial oxidative phosphorylation; on the other hand, there is vitamin K, another physiological quinone that exerts its antihemorrhagic effect by accelerating tissue oxidation and favoring clotting. In excessive amount, vitamin K causes neoplastic degeneration by an excessive production of free radicals. And there is a whole series of other quinones that accelerate cell oxidation in an abnormal manner; again, they are particularly dangerous because of the consequent production of large amounts of free radicals. Polluted air, for example, is very rich in quinone substances, and most of the damage caused to the lungs by atmospheric pollution is due to the high level of oxidative stress induced by these substances, which have a proinflammatory and cytotoxic effect. However, it has been demonstrated that if a low level of oxidative stress is induced (e.g., with small homotoxicological amounts of quinone), antioxidant defenses are instead activated in the cell.

Quinones not only trigger production of a large amount of free radicals in the tissues but are also capable of binding irreversibly with some protein amino acids, provoking the inactivation of the proteins themselves and the loss of their function. For this reason, quinone substances are widely used as antimicrobial, antiparasitic, and antineoplastic agents. Many anthraquinones in common medicinal plants that we use for other purposes, such as *Cassia* and *Hypericum*, also have powerful antimicrobial activity precisely because of these biochemical considerations.

In order to understand in depth the role of quinones in cell metabolism and the possible use of small amounts of these substances in therapy, it is necessary to take a step backward and briefly analyze the biochemical process that develops in the mitochondrion and leads to the use of oxygen and the production of energy, known as oxidative phosphorylation.

The NADH and $FADH_2$ formed during the Krebs cycle and the oxidation of fatty acids in the mitochondrial matrix are molecules rich in energy because they contain a pair of electrons with a high transfer potential. When these electrons are used to reduce molecular oxygen to water, a large amount of energy can be used to generate ATP. Oxidative phosphorylation is the biochemical process by which ATP is formed as a consequence of the transfer of electrons from NADH or $FADH_2$ to molecular oxygen by a series of electron carriers (see **Fig. 4.40**). This process is the largest source of energy for aerobic organism cells.

This electron transfer through these localized protein complexes in the inner membrane of the mitochondrion starts the pumping of protons to the outside of the mitochondrial matrix (see **Fig. 4.40**). The resulting proton redistribution generates a pH gradient and a transmembrane electric potential that is the proton-motive force. ATP is

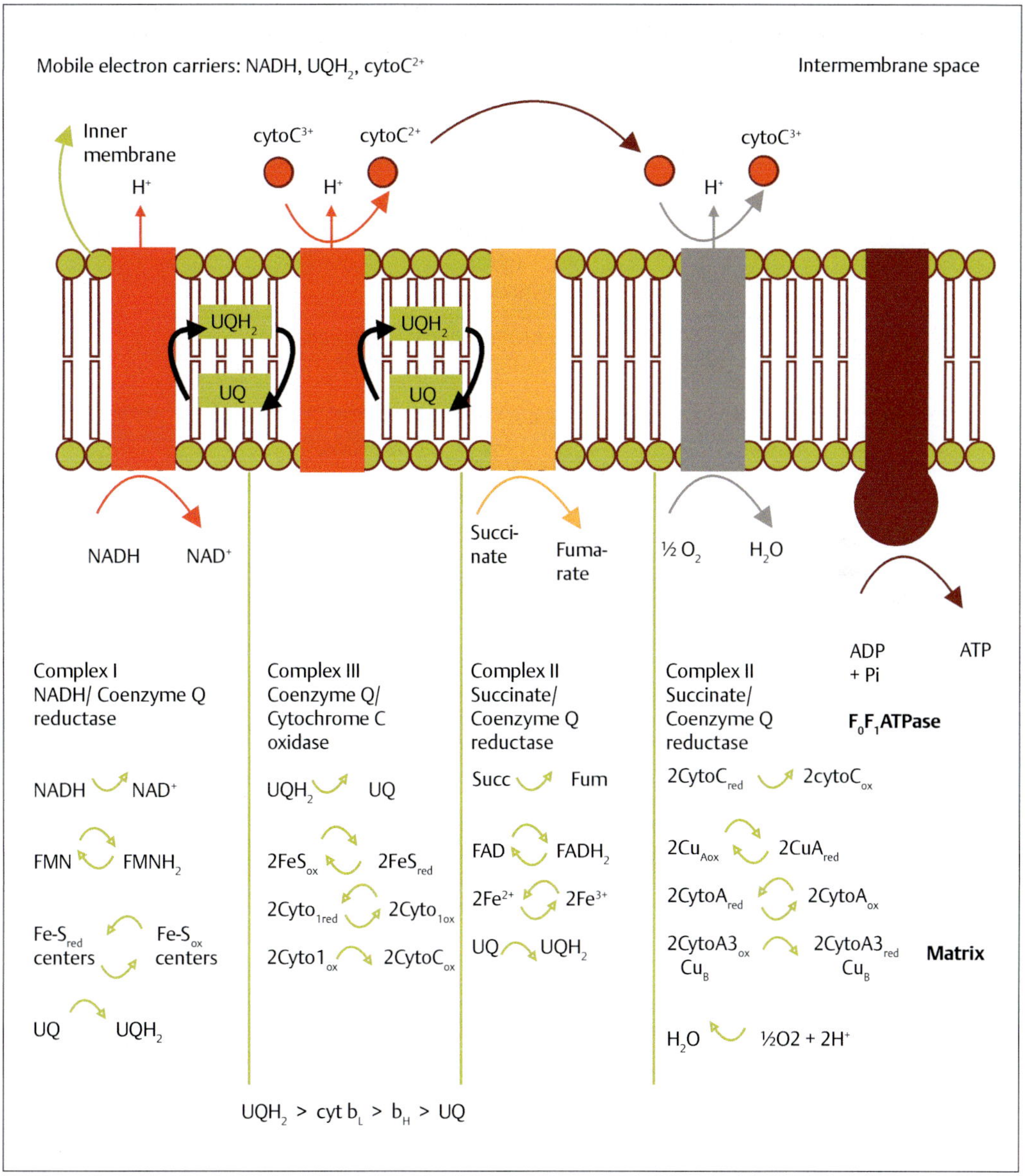


**Fig. 4.40** Mitochondrial electron transport.

synthesized when the protons flow back into the mitochondrial matrix through a whole series of enzyme complexes. Oxidative phosphorylation is, therefore, the culmination of a whole series of enzymatic transformations that, taken as a whole, represent cell respiration. Initially, the fuel—glucose, amino acids, and fats—is oxidized in the citric acid cycle to store electrons with a high transfer potential. Then, this electron motive force is converted into a proton motive force that finally is converted into a potential capable of binding the phosphorus atom in ATP.

The institution of this energy storage process in the phosphate radical was the moment that started life on earth because it supplied the first storable fuel for living matter.

The transformation of electron energy into proton energy is catalyzed by three pumps: (1) NADH-Q oxidoreductase, (2) Q-cytochrome-*c* oxidoreductase, and (3) cytochrome-*c* reductase (**Fig. 4.41**). These large transmembrane complexes contain multiple redox centers in which there are quinones, flavones, iron-sulfur clusters, hemes, and copper ions. The reactions, from glycolysis to

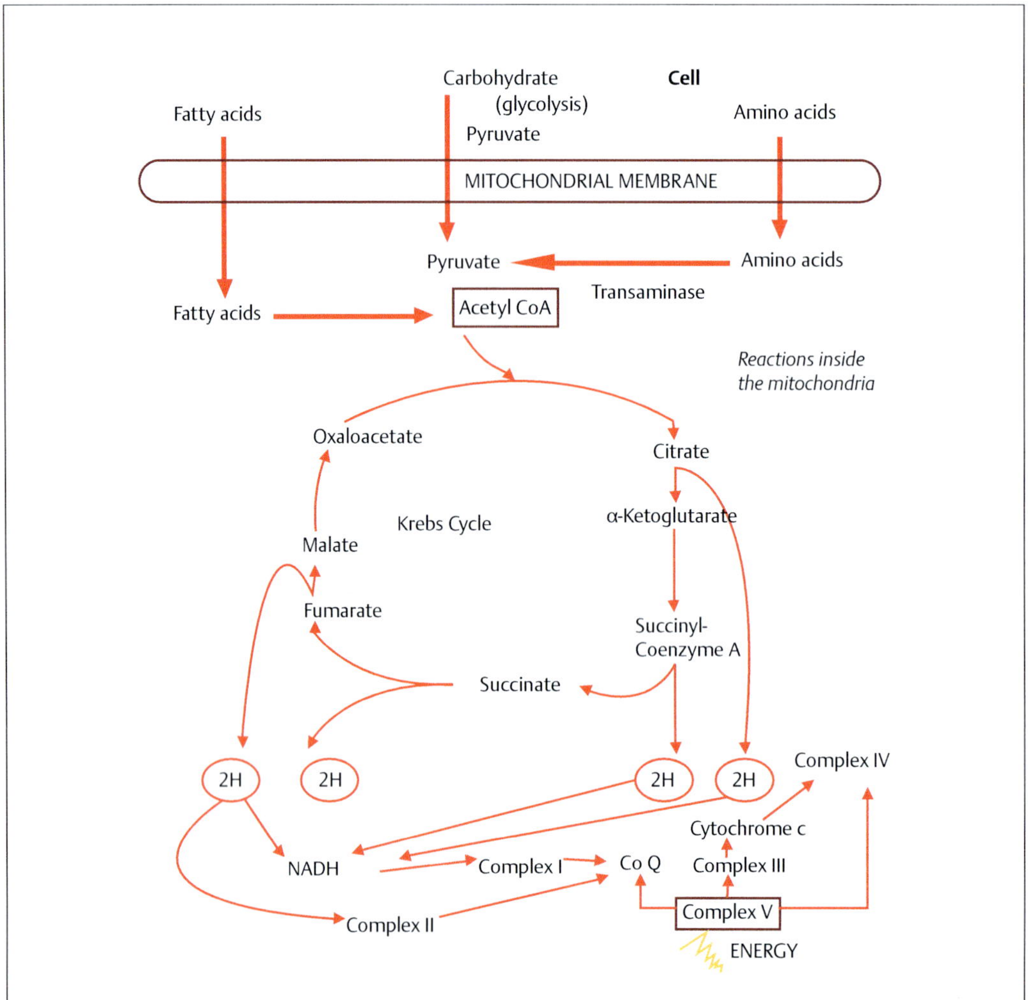


**Fig. 4.41** The transformation of electron energy into proton energy.

the formation of ATP in oxidative phosphorylation, occur within about 1 millisecond, which means that the two processes cannot be separately influenced. Therefore, when we want to influence mitochondrial energy production with homotoxicological products, we must simultaneously intervene in the Krebs cycle with intermediate catalysts and in oxidative phosphorylation with the appropriate quinone.

The speed of these reactions, the consequent use of oxygen, and the consequent production of free radicals depend on the redox potential of molecules such as quinones. A proportion of free radicals is also produced physiologically in oxidative phosphorylation using endogenous ubiquinone, and this to a certain extent always damages the mitochondrial structure. Cell respiration is both a vital and a lethal process: on the one hand, energy for each vital function is produced; and on the other, free radicals that progressively destroy the cell structures are also produced (**Fig. 4.42**). If the energy functions of a cell or a body are stimulated, as has been tested recently in *Drosophila*, then life expectancy decreases accordingly.

However, oxidative phosphorylation is an extremely fine and delicate mechanism that must not just be stimulated, as is done with allopathic methods by administering large amounts of coenzyme Q, but rebalanced by administering potency chords of physiological quinone wherever a slowing down in the cell respiratory functions is expected. A whole series of very frequent pathological conditions, usually linked to bodily aging, is in the final analysis ascribable to a slowing down of the process of oxidative phosphorylation, which leads to a reduction of the metabolism and to the appearance of problems such as hypercholesterolemia, arteriosclerosis, hypertension, and depression. Ubiquinone in combined homeopathic dilutions is a specific cell metabolism rebalancing drug.

There are, however, some situations in which physiological slowing with age is not to blame; mitochondrial damage may be linked to exogenous toxins. In these cases, we must use small homotoxicological doses of corresponding quinones, which will antagonize excessive or defective oxidation of these anomalous quinones.

Therefore, naphthoquinone, in various combined dilutions and dynamic formulations, will be indicated in relation to pollution in the patient with chronic bronchitis. Anthraquinone will be

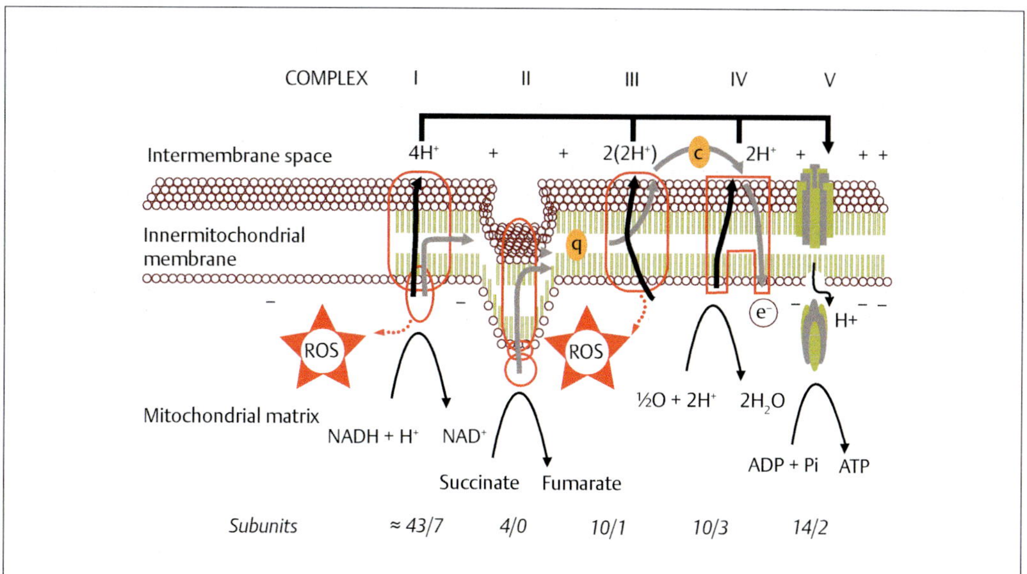


**Fig. 4.42** Free radicals (ROS, reactive oxygen species) that progressively destroy the cell structures are inevitably produced during cell respiration.

indicated in the same form in patients intoxicated with laxatives, and so forth[70].

### The Role of Ubiquinone in the Respiratory Chain

Quinones play a vital role in cell function. Their role is that of creating, through separation of electrons and protons of hydrogen atoms, a potential difference at the inner membrane of the mitochondrion of at least 220 mV, a level above which there is energy available to add a phosphorus atom to ADP. Molecular oxygen is the final electronic acceptor of the chain, and the series of mitochondrial energy reactions stops in the absence of a sufficient amount of oxygen. The crucial importance of the supplementation of elementary oxygen, particularly in the prevention and care of chronic and degenerative diseases, will be detailed in another section. There is scientific evidence according to which aberrant cells, particularly cancer cells, become so because of the nonavailability, albeit intermittent, of oxygen, which brings about an anaerobic type of metabolic state and activates the Embden–Meyerhof cycle, typical of the neoplastic cell, to be replenished with ATP. One way to find out whether cells are assuming these characteristics is to measure their consumption of glucose, because a neoplastic cell uses three times more glucose than normal.

Ubiquinone is present in considerably larger amounts than cytochrome *c* during oxidative phosphorylation; in addition, its electron transport capacity is twice that of the cytochrome itself, so that ubiquinone has a preponderant role in mitochondrial energy production. In order to provide an idea of amounts, we should recall that for each molecule of NADH dehydrogenase, there are 3 molecules of cytochrome $bc_1$, 7 molecules of cytochrome oxidase, 9 molecules of cytochrome *c*, and as many as 50 molecules of ubiquinone in the mitochondrion.

The sequence of oxidative phosphorylation transfers goes in a single direction as it follows an energy gradient that increases gradually as the oxygen atom is approached. We should, however, remember that ATP synthesis is a reversible reaction that can go in the ADP + Pi = ATP direction when there is a sufficient energy gradient or in the opposite direction to pump hydrogen protons if the energy gradient is reduced.

In the neoplastic cell, this ordered sequence of chemical reactions is inverted in connection, usually, with a viral or toxic problem, in accordance with a not yet completely clarified mechanism. In short, the cell undergoing neoplastic transformation is restricted to converting its production of energy in an anaerobic direction and forms, as was observed by Warburg in 1920, a large amount of lactic acid—124 times as much as a normal cell. Basically, a neoplastic cell removes 70 mg of glucose from 100 mL of blood, returning 46 mg of lactic acid.

All these ideas are important in understanding the etiopathogenesis of cancer at the cellular level and in organizing a reasonable care strategy. Currently, the neoplastic transformation is simply attributed to the action of carcinogens in the cell; however, it appears more plausible to attribute this degenerative thrust to the functional alteration of the mitochondrial respiratory chain. The carcinogenic substance integrated in the cytochrome $bc_1$ oxidase complex or in another part of the oxidative chain prevents ubiquinone from developing its fundamental role of electron donor. The possibility that this anomalous integration takes place in the mitochondrion can be attributed to the temporary anoxia that, for various reasons, attacks the cell. The role of the scarcity of oxygen in carcinogenesis is, therefore, primary. These notions are based on rigorous tests, performed by Goldblatt and Cameron in the 1950 s[71,72].

According to the homotoxicological view, the cancer problem cannot be attributed solely to the appearance of mutations induced in the cell by carcinogens and viruses. This explains why not all heavy smokers contract cancer and why not all those who come into contact with asbestos particles develop pleural mesothelioma.

It is calculated that at least $10^{16}$ genetic mutations occur during the course of a human life, and even in a protected environment mutations occur

---

70 Bianchi I. Geriatria ed omotossicologia. *Rivista Italiana di Omotossicologia.* 1989;(1):15–24

71 Goldblatt H et al. Induced malignancy in cells from rat myocardium subjected to intermittent anaerobiosis during long propagation in vitro. J Exp Med. 1953;97(4):525–552

72 Bianchi I. Geriatria ed omotossicologia. *Rivista Italiana di Omotossicologia.* 1989;(1):15–24

spontaneously at a rate of $10^{-6}$ per gene for each cell division. During a lifetime, every single gene presumably experiences $10^{10}$ mutations; this is a guarantee of adaptability to the environment and brings chances of survival to the species. Upon assessing these data, we may naturally consider that the incidence of cancer, in the living, is fairly rare compared with the level of genetic changes. Moreover, if we ascribe cancer to a genetic etiology alone, this ought to be relatively independent of age, whereas epidemiologists demonstrate that this pathological condition is closely related to the age of the individual. In the etiopathogenesis of cancer, factors that go beyond simple genetic damage play a fundamental role. For the four most common types of cancer (prostate, breast, colorectal, and lung), a role for scarcity of cell oxygen can be envisaged. If the problem of the neoplastic cell is a metabolic change, a revolutionary therapeutic suggestion would be that of not destroying damaged cells, as is usually done in conventional medicine, but of correcting metabolic cell damage (e.g., by forcing a blockage of the oxidative chain using a potentiated and much more active form of quinone, such as parabenzoquinone). A clinical trial based on these notions and performed using parabenzoquinone in 66 patients with colorectal cancer has had considerable success. In addition to an improvement in quality of life, there were significantly fewer metastases in patients who underwent this treatment compared with control subjects[73].

**Important Quinones in Clinical Medicine**

In the following will review the various quinones in detail in order to understand their specific therapeutic role and to learn their use not only in ponderal form but also in small homeopathic-homotoxicological doses.

From the therapeutic viewpoint, quinones can be used in three different forms with three different groups of overlapping indications:

- When we want to use quinones in an antibiotic or antineoplastic role, they are used in ponderal doses (chemotherapeutic).
- When we want to use quinones as a stimulus for cell reparative processes, they are used in infinitesimal doses (Ubichinon compositum).
- When we want to use quinones as a stimulus to the rebalancing of mitochondrial oxidative phosphorylation, which changes progressively with age, they are used in homotoxicological potency chords.

Quinones occur in great profusion in nature. The colors of the leaves in the fall reflect the accumulation of specific quinones, and many medicinal plants act precisely because of their quinone content. Quinones are also often derived from tar and many chemical substances commonly used by modern industry. Quinone substances are often carcinogenic because they induce excessive tissue oxidation and, therefore, the production of large amounts of free radicals. The effect of quinone accumulation is first clogging and then blocking of mitochondrial function. This can also be verified clinically because where there is quinone accumulation there is melanosis and tissue browning. Where there is quinone accumulation, there is a deficient consumption of oxygen and, therefore, a local decline in temperature. One of the first signs of effective quinone therapy will be a local or general increase in temperature, always to be seen as an extremely positive event of functional and reactive recovery.

The homotoxicological idea of the use of quinones arises in part because of homeopathy, but also because of considerations of biochemistry and the Arndt–Schulz rule. Small amounts of a very active and sometimes toxic substance have a reactive stimulus effect and antagonize the blockage that ponderal amounts of this substance have caused. Quinone therapy must be combined with correct and deep drainage, a toxin-free diet, and psychoemotive rebalancing.

Quinones have a proven effect on the neoplastic cell. This effect is amplified by combination with ascorbate and is in direct relation to their redox potential. In addition, quinones have a marked anti-infective ability; however, this is selective and related to their chemical structure. Quinones with an *ortho* structure have a more marked antibacterial effect, whereas quinones with a para structure have a more marked antimycotic effect.

---

[73] Jakab F et al. A medical nutriment has supportive value in the treatment of colorectal cancer. *Br J Cancer.* 2003;89: 465–469

### Quinones in Homotoxicology

Homeopathized quinones used in individual combinations of potencies or as single dilutions within the ambit of a compound drug are a fundamental contribution of homotoxicology to homeopathy. With the introduction of these drugs, it is possible to use substances in infinitesimal dilutions that are harmless to cell respiration. Most of the basic ideas on the use of these diluted substances are described in the original works of William F. Koch, in the studies and tests of Sigrun Seutemann and Raymond Kastner, in the tests of the physicians of the Italian School of Homotoxicology, and in studies of the biochemistry of these substances.

Quinones have different clinical uses according to their structures and, therefore, the type of binding energy that they have. We must always bear in mind the redox potential of individual quinones (see **Fig. 4.39**). The higher this potential is, the more stongly oxidant the quinone is and the deeper its action is. Etiological multifactoriality, typical of most pathological conditions, justifies the combined use of various quinones. Each of them, however, has its own characteristics, profundity of action, and organic tropism. Accordingly, the formulation of well-founded therapeutic combinations requires the study of individual quinones[74].

### William F. Koch and Hans-Heinrich Reckeweg

Just as Reckeweg's source for the citric acid cycle intermediate catalysts was the discoveries of Krebs, so for quinones and glyoxals the source of information leading to the inclusion of these homeopathically diluted products in homotoxicological therapy was the studies and passionate research of a US physician, William F. Koch (1885–1967).

Koch was born in Detroit, Michigan, in 1885. A physician and a chemistry, histology, and physiology professor, he invented and patented chemical substances capable of reactivating cell oxidative mechanisms, allowing the body to initiate a self-healing process. After his first few years of university work, Koch became interested in problems connected with intoxication, both exogenous and endogenous, often induced by particular diseases. His first studies concentrated on the function of the parathyroid glands, whose anabolic role he recognized for the first time. In addition, he stated that tetanic symptoms encountered in patients who underwent parathyroidectomy were not just due to hypocalcemia but were also related to an intoxicatory state that was noted in these patients. He also noted that the urine of patients and animals that underwent parathyroidectomy contained large amounts of lactic acid as proof of poor functioning of the cell oxidative processes in the mitochondrion.

Subsequently, he noted that tissues with a greater concentration of carbonyl groups had more efficient oxidative processes and that they were more resistant to cancer, characterized particularly by blocking of the oxidative metabolism and of the induction of consequent tissue acidosis. Excessive amounts of guanidine, methylguanidine, dimethylguanidine, and other alkylated substances such as methylcyanamide, responsible for tetanic symptoms and the death of these subjects, were also present in the urine of patients who underwent parathyroidectomy. According to Koch, the parathyroid glands were in some way responsible for the purification of these substances, which could not be isolated in the tissues of healthy subjects. Koch prepared a type of serum that proved to be very effective in neoplastic subjects. After the injection of the serum, no symptoms initially occurred; however, after about 24 hours, fever, nausea, vomiting, depression, and various focal symptoms appeared and were interpreted to be a consequence of necrosis of the neoplastic mass. The severity and duration of the symptoms were thus correlated with the stage and size of the tumor.

Koch's basic clinical notion was very close to that of Reckeweg's later notion. He thought that neoplasia was also in some way a defense reaction, although insufficient and inadequate, by the body against external toxic or microbial attack. Although Koch's formula is unknown, we do know that he studied several substances, in particular quinones and aldehydes, which, injected in minute doses, should have an oxidative stimulus effect on the body. In fact, it is the defective functioning of the Krebs cycle (mainly of oxidative phosphorylation) that, because of the poor use of oxygen and production of energy, allows the onset of infective and neoplastic pathological conditions. Thus, Koch's therapy, suitable for reactivating a physio-

74 Bianchi I. *Geriatrics and Homotoxicology*. Baden-Baden: Aurelia Verlag; 1994

logical mechanism, was understood as profoundly etiological.

The chronology of the events coincides perfectly with the introduction of these new elements of homotoxicological pharmacology, which was also favored by the studies of Seutemann. Seutemann was an expert in biochemistry as applied to homeopathy and worked in the 1970s in close collaboration with Hans Reckeweg and Reinhard Voll.

The most important substance described by Koch was called "glyoxylide"; however, this specific formula has been lost. Koch and his collaborators treated many cases of cancer, tuberculosis, and other complex pathological conditions with this formula and other members of the quinone family. They treated more than 20000 patients, with a mean cure rate of 46% in the neoplastic field. In 1919, the head of the American Medical Association, Morris Fishbein, tried to acquire and apply Koch's method, without success. Afterward, Fishbein did everything he could to prevent clinical and pharmacological research with these preparations anywhere in the United States. Koch was cited in a Food and Drug Administration judgment but was acquitted after 600 doctors gave evidence in his favor. Koch died of poisoning on December 9, 1967.

Koch's studies mainly concentrated on identifying the possibility of inducing regression of a pathological condition even when the condition was chronic. This was not without interest to Reckeweg, who in the same years talked about the importance of obtaining with the therapy a regressive vicariation that would lead to recovery, even from chronic degenerative diseases. According to Koch, we should concentrate on the cell oxidative mechanism and on the consumption of oxygen, which in viral and neoplastic pathological conditions was almost always blocked. Koch was convinced, as was Reckeweg, that the body has an innate ability to recover and maintain health, even in fairly serious diseases such as cancer. Within the ambit of this research, Koch discovered and developed many synthetic antitoxins, including glyoxylide, malonide, and parabenzoquinone. These catalysts, in his opinion, gave the cell the necessary oxidative stimulus to cut the link that in chronic diseases is created (Reckeweg's homotoxone) between the host cell and the pathogen. This pathogen could be a virus, a chemical carcinogen, a bacterial toxin, or an incompletely oxidized tissue metabolite. He identified the toxic free radical in the amine group of the toxin that binds to the quinone carbonyl. The solution was to give energy to the cell to cut this link, by stimulating mitochondrial oxidative phosphorylation. He was, thus, going in the opposite direction to conventional medicine, which from then on proclaimed surgery, radiotherapy, and later chemotherapy as the only therapeutic possibilities in chronic pathological conditions.

Conventional medicine launched a denigrating campaign, lasting more than 50 years, against Koch and his destabilizing theories. From the 1920s onward, anyone who attributed chronic diseases to environmental toxins, vaccinations, dietary deficits, and oxidative deficits in the cell would appear unscientific. Unfortunately, Koch took revenge by leaving no traces of how to synthesize his more complex drugs; however, quinones, glyoxal, and other remedies are today an extremely important inheritance of homotoxicology.

*The Quinones*

### Hydroquinone

#### *Ponderal Doses*

A ponderal dose of hydroquinone is obtained by reduction of parabenzoquinone. In fact, hydroquinone is not a true quinone because the carbonyl group no longer exists, being replaced by two –OH groups (**Fig. 4.43**). However, it is found in the body as an oxidative phosphorylation intermediate and is immediately oxidized by cytochrome *c*. Hydroquinone has marked reducing characteristics and is used for this in photography. It is used as an antiseptic and disinfectant in medicine because of these same properties.

#### *Homotoxicological Doses*

The marked reducing ability of hydroquinone makes it an excellent antipyretic, antibacterial, and disinfectant. In urinary infections with alkalinic urine, hydroquinone is very effective; moreover, everyone is familiar with its efficacy in such clinical situations of tincture or infusion of *Uva ÿuinin, a plant rich in hydroquinone.*

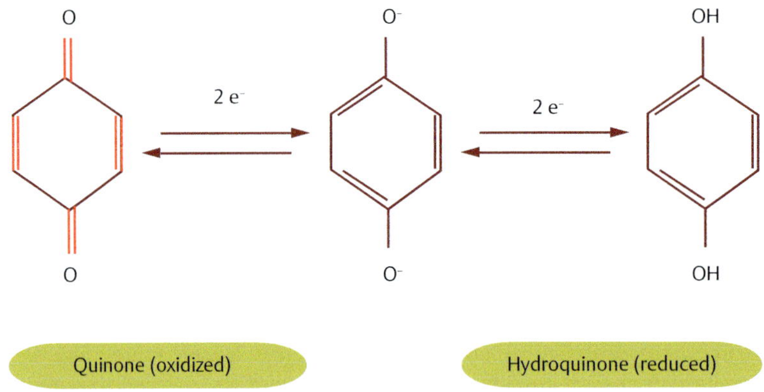


**Fig. 4.43** Hydroquinone.

Hydroquinone in homotoxicological form is used in the initial phases of uinine therapy, so as not to start with too active a remedy. On the basis of clinical experience there are specific indications for the use of this uinine, which include the following:

- Psychophysical asthenia
- Neuralgia
- Accumulation of toxins
- Damage from alcohol, smoking, coffee, and drugs
- Consequences of miscarriage
- Depression
- Psoriasis
- Chronic inflammations
- Precancerous conditions

### Quinhydrone

#### *Ponderal Doses*

Quinhydrone is an addition compound that contains one molecule of parabenzoquinone and one molecule of hydroquinone (**Fig. 4.44**). Wohler coined the term *quinhydrone*, which combines the two components, in 1844. Quinhydrone forms needlelike red-brown crystals with a green surface. It is not very soluble in cold water, but it is

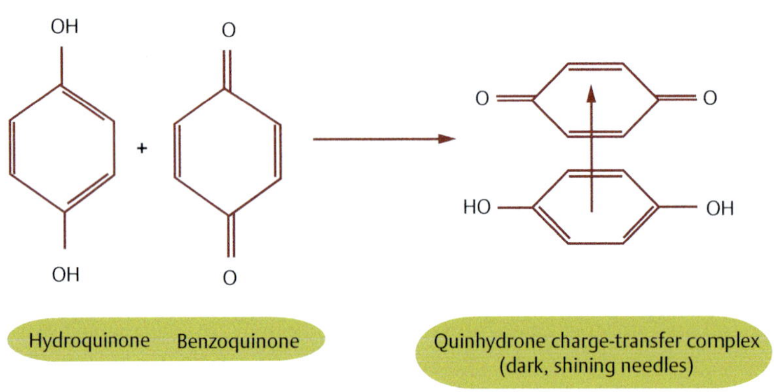


**Fig. 4.44** Quinhydrone.

easily soluble in ether, alcohol, and hot water. In solution, it dissolves until it reaches an equilibrium, behaving like an equimolecular mixture of uinine and hydroquinone. Quinhydrone is used in special electrodes for potentiometric measurement of the pH level in a solution.

Quinhydrone plays a decisive role in the intermediate metabolism of hydroquinone and para-benzoquinone. Therefore, by not being a potent acceptor of hydrogen, it balances the stimulation function of oxidative phosphorylation performed generally by quinones.

### *Homotoxicological Doses*

Quinhydrone is a complementary quinone and is less active than the others covered herein. It is particularly useful when organic reactivity is blocked by allopathic drugs, viral pathological conditions, heavy-metal intoxication, geopathic stresses, and pollution. It has an effect of reactivation and drainage and acts best when associated with other catalysts, such as Natrium pyruvicum, and trace elements, such as manganese, nickel, and molybdenum in particular.

## Anthraquinone

### *Ponderal Doses*

Anthraquinone (**Fig. 4.45**) is obtained by oxidation of anthracene, a tar-colored substance. Anthraquinone forms odorless rhomboidal crystals whose color varies from yellow-green to yellow. This substance is insoluble in water, soluble with difficulty in ether and alcohol, and easily soluble in hot benzene. The redox potential is 154 mV; its oxidizing power is therefore less than that of the other quinones mentioned. This compound is found in almost all plants classically used as laxatives (rhubarb, senna, cascara, and aloe) and in many synthetic laxatives. It is a starting molecule for the preparation of numerous dyes; alizarin, a natural red-purple dye, is an anthraquinone derivative. An anthraquinone derivative is used in agriculture as a denaturant. It is this anthraquinone that gives red color to cereal seeds, vegetables, and greenery in general, which is useful for making them less palatable to birds.

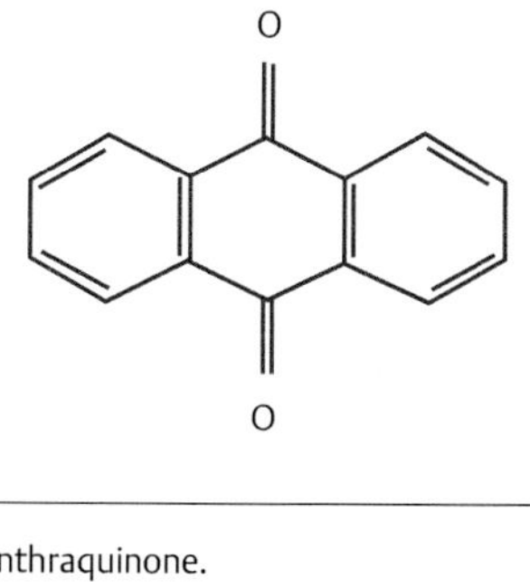

**Fig. 4.45** Anthraquinone.

Plants containing anthraquinone (aloe has the greatest percentage) act specifically in the large intestine by stimulating peristalsis and retaining water in the intestinal lumen.

### *Homotoxicological Doses*

Koch believed that anthraquinone in diluted and dynamized form stimulates toxin combustion in the cell, regenerating the mitochondrial enzyme system and reactivating oxidative phosphorylation. Therefore, anthraquinone or compounds derived from it are widely used today. From synthetic drugs to herbal laxatives and agricultural treatment residues that stay in food, our bodies receive an overload of anthraquinone substances. It follows that the use of anthraquinone in homotoxicological unblocking form is, and must be, very frequent. However, toxin mobilization in the patient, particularly if the patient is elderly and intoxicated, must be performed very cautiously. Intense drainage through eliminatory organs must be performed consecutively, in order to avoid toxins, which are deposited in the various tissues of mesodermal origin. These toxins cause painful symptoms or pathological conditions of a reactive inflammatory type.

With the use of anthraquinone in homotoxicological doses, the lymphatic system and immunological reactivity can be stimulated, with benefits for the skin and the mind. The most frequent clinical indications treated according to the *Homeopathic Homotoxicology Repertory & Materia Medica* include:

- Hypochondria
- Manias
- Skin pathologies in general
- Digestive system disorders
- Venous insufficiency

- Consequences of laxative abuse
- Drug intoxication
- Rectal and precancerous conditions
- Digestive melanosis
- Psychic hypersensitivity
- Thyroid hyperfunction

### Naphthoquinone

#### *Ponderal Doses*

Naphthoquinone (**Fig. 4.46**) is obtained by oxidation of naphthalene. It occurs usually in the form of sharp, irritating needles that are intense yellow to orange in color. It is not very soluble in water, but it is easily soluble in ether, alcohol, benzene, and chloroform. It is volatile in steam. The redox potential is 484 mV.

Naphthoquinone is an essential component of vitamin K (**Fig. 4.47**). In nature, there are more than 120 naphthoquinones. Naphthoquinones are present in many plants that are commonly used, such as those of the genera *Juglans* and *Drosera* in particular.

Naphthoquinone is used in industry as an antioxidant. It is used mainly as a polymerization regulator in the production of synthetic gum and polyester resins. It is also used to prevent the proliferation of algae and as an antimycotic agent in medicine.

According to Koch, naphthoquinone has a tropism specific to the whole endocrine system, including the pancreas. It is indicated in thyroid hypofunction. It also has a specific action on the lymph glands, including the tonsils and the appendix. Naphthoquinone seems to have a modulating action on the parasympathetic-vagal system.

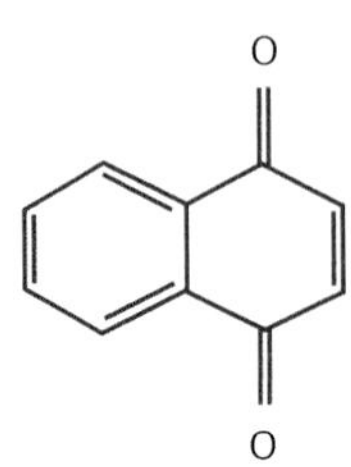


**Fig. 4.46** Naphthoquinone.

#### *Homotoxicological Doses*

On the basis of the previous considerations and the experience of Koch and his pupils and the Homotoxicological School, there are extensive clinical indications for the use of this quinone in homotoxicological form. These indications are analytically reported in the *Homeopathic Homotoxicology Repertory & Materia Medica*. The most relevant indications include the following:

- Care and prevention of digestive (colon) and genital neoplasias
- Intestinal degenerative pathologies
- Prevention and treatment of prostate cancer
- Consequences of radiotherapy
- Thyroid hypofunction
- Prolonged fever and weight loss of unknown origin
- Pathologies tending to hemorrhage

### Parabenzoquinone

#### *Ponderal Doses*

Parabenzoquinone (**Fig. 4.48**), a derivative of the oxidation of hydroquinone, was discovered in 1838 by Woskressenky as a derivative of quinic acid, a component of cinchona bark. It irritates the mucosae and colors the skin brown, causing necrosis by long contact. It is used in industry to produce dyes, in photography and the tanning industry for its powerful oxidant properties, and as a reagent in analytical laboratories.

Parabenzoquinone has a redox potential of 0.715 V, a molecular weight of 108.09, and an infrared wavelength of 1669 $cm^{-1}$. It forms yellow-gold crystals with a pungent odor similar to that of chlorine. It is not very soluble in water, but it is soluble in alcohol, ether, petroleum, and alkaline solutions.

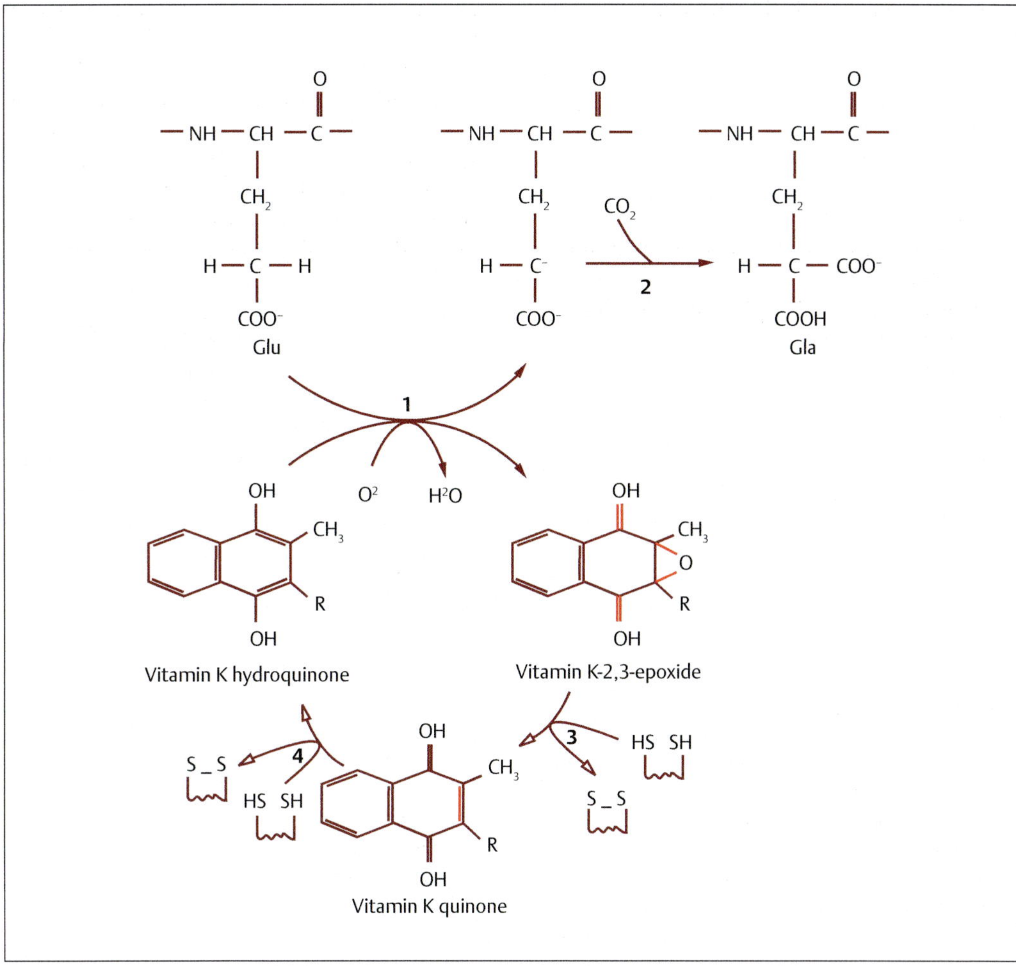


**Fig. 4.47** Naphthoquinone and vitamin K.

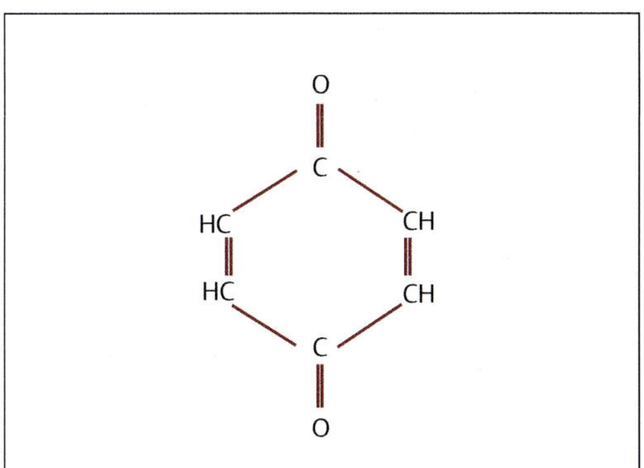


**Fig. 4.48** Parabenzoquinone.

### *Homotoxicological Doses*

According to Koch's studies, parabenzoquinone has a specific tropism to the adrenal glands and this, therefore, needs to be taken into account when prescribing it. Thus, it is an excellent tonic and can be used as a palliative treatment in subjects affected by degenerative or neoplastic pathological conditions, in whom it in general improves the energy level. Detailed clinical indications for the use of this substance in homotoxicological form are reported in the *Homeopathic Homotoxicology Repertory & Materia Medica*. The clinical conditions that can most benefit from the administration of this substance diluted in homotoxicological form include the following:

- Nervous stress
- Intellectual deficit
- Mental torpor
- Eczema
- Anorexia
- Sequelae of viral and venereal diseases and vaccinations
- Drug, medicine, and smoking intoxications
- Pancreatic insufficiency
- Proliferating tendency

### Triquinoyl

*Ponderal Doses*

Triquinoyl (**Fig. 4.49**) is the product of oxidation of hexahydroxybenzene: it forms colorless needles; is almost insoluble in cold water, alcohol, and ether; and is soluble in alkali and alkaline earth salt solutions. In alkaline solution, it decomposes into $CO_2$ and rhodizonic acid. Triquinoyl is an extremely powerful quinone with 6 carbonyl (C=O) groups that are all free (i.e., active in the benzene ring). These groups function as strong electron donors without fixing free radicals. Therefore, this substance can substantially activate and possibly unblock mitochondrial electron transport. The resumption of the use of oxygen and the consequent correct production of large amounts of energy allow the cell to produce interferons to antagonize viral attacks and to effectively repair DNA damage. Triquinoyl is one of the most energetic electron carriers used in biological medicine.

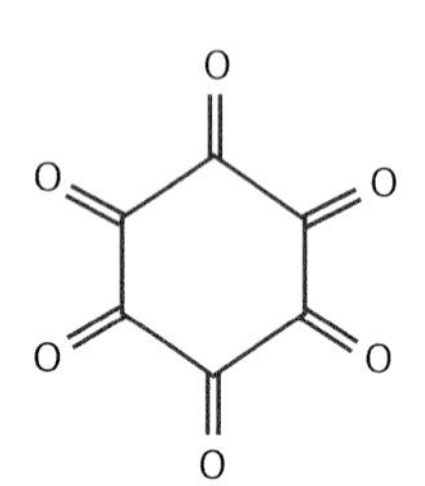

**Fig. 4.49** Triquinoyl.

Triquinoyl acts in a practically identical manner to the famous "glyoxylide" (O=C=C=O), a compound that Koch prepared by dehydration of glyoxylic acid (CHO-COOH). It is this substance that the famous physician used with success in numerous pathological conditions, including diabetes, arthritis, poliomyelitis, and even cancer.

*Homotoxicological Doses*

Triquinoyl, rhodizonic acid, and their polymers are fundamental elements of homotoxicological therapy. Being extremely powerful substances, their ideal use is precisely in a diluted and dynamized form. Many physicians have used this homeopathized substance in a large number of chronic pathological conditions. Seutemann mentions obtaining hundreds of regressive vicariations, in particular in patients affected by lymphoma, with combinations of hydroquinone, parabenzoquinone, and triquinoyl. The clinical indications of these administrations are extensively reported in the *Homeopathic Homotoxicology Repertory & Materia Medica*. The pathological situations in which homotoxicological triquinoyl is indicated include the following:

- Depression
- Weight loss
- Viral hepatitis
- Intestinal polyposis
- Anemia
- Postinfective asthenia
- Adenomegaly

### Ubiquinone

*Ponderal Doses*

Coenzyme $Q_{10}$ is also called ubiquinone because it is present with similar chemical structures in both animal and plant cells (**Fig. 4.50**). It is the only component of the mitochondrial electron transport chain not bound to protein. The long hydrophobic side-chain of 10 isoprenoid units gives coenzyme $Q_{10}$ excellent solubility in lipids, which allows it to diffuse through the internal mitochondrial membrane. When the oxidized form of the quinone accepts the first electron, it forms a free radical that extremely dangerous to the mitochondrial and cell structures. At this point, the mole-

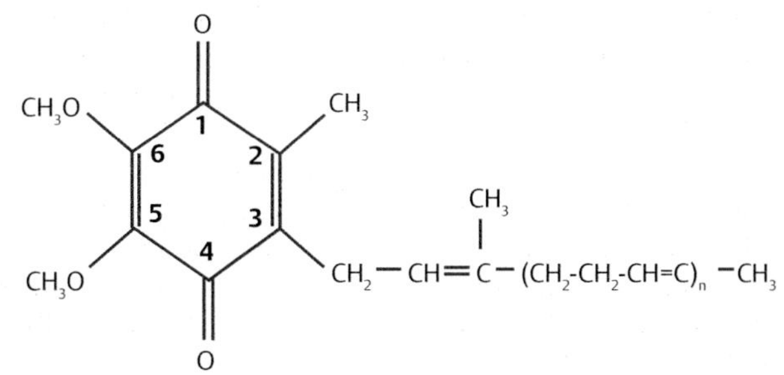

**Fig. 4.50** Ubiquinone.

cule can accept a second electron and two protons to form a completely reduced and innocuous quinone. At the same time, proton displacements and ATP synthesis take place.

### *Homotoxicological Doses*

It is understood that the correct functioning of ubiquinone is fundamental for a good level of cell energy production necessary for each cell function, from protein synthesis to regulation of proliferation or cell apoptosis, fundamental processes in the immunological field and in the control of neoplastic proliferations. Along with the aging of the body, or in the course of metabolic problems, the function of ubiquinone can encounter obstacles that have profound repercussions on cell and organ function. In our opinion, the much-vaunted supplementation of ubiquinone is not correct; by this we increase the process of mitochondrial oxidation, the production of energy, and the formation of free radicals, which will inevitably tend to destroy the delicate mitochondrial DNA. Instead, it is sensible to have supplementation-regulation, as can be obtained by administering infinitesimal dynamized doses of ubiquinone in various dilutions, which is typical of homotoxicology. The clinical indications for these administrations are reported in detail in the *Homeopathic Homotoxicology Repertory & Materia Medica*. The pathological situations in which homotoxicological ubiquinone is indicated include the following:

- Intellectual deficit
- Irritability
- Depression
- Retinopathy
- Learning difficulties
- Dyslipidemia
- Tendency to weight gain
- Elderly endocrine disorders
- Immunological deficit
- Medullary deficit
- Bronchitis sequelae
- Hepatic steatosis
- Arterial hypertension
- Atherosclerosis
- Muscular atrophy

#### *Practical Application of Homotoxicological Therapy to Changes in Oxidative Phosphorylation*

Despite our efforts and those of many other researchers and clinicians, it is clear, from these pages and even more from a detailed study of the section dedicated to quinones in the *Homeopathic Homotoxicology Repertory & Materia Medica*, that the clinical indications of the various quinones are in part overlapping. It is not easy, in contrast to the Krebs cycle catalysts, to make a clear distinction in clinical indication between the various quinones. This is not always the case because there are peculiarities and pathognomonic symptoms for each quinone, such as the indications of anthraquinone in hyperthyroidism and of naphthoquinone in hypothyroidism. There is an organ tropism of anthraquinone at the skin level and ubiquinone at the metabolic level. However, comprehensive assessment whether all quinones have a common oxidizing activity at the cellular level shows that all stimulate the correct expression of mitochondrial oxidative phosphorylation; all have a key role in the prevention and care of degenerative pathological conditions. It is, therefore, sensible to employ Reckeweg's idea of grouping together various quinones, in appropriate dilution, in a single drug for antidegenerative and general activating quinone stimulation for simple implementation without recourse to the use of bioelectronic devices or the in-depth study of the *Homeopathic Homotoxicology Repertory & Materia Medica.*

Ubiquinone compositum is one of the most important and effective homotoxicological drugs, indicated for the reactivation of cell function, the consumption of oxygen, the production of energy, and the antagonism of cellular degeneration episodes, thanks to its composition of phytotherapeutic-homeopathic substances with an antineoplastic effect.

### *Concluding Summary on Quinones*

- Quinones are a group of natural or synthetic chemical substances that have as a common element an outstanding ability to bind oxygen.
- Some of them (e.g., ubiquinone) are physiologically present in the mitochondrion and, together with cytochromes, dehydrogenases, and flavoproteins, play a fundamental role in the transport chain that conveys electrons (the same electrons that arise in the form of hydrogen atoms from the Krebs cycle) toward molecular oxygen.
- However, toxic quinones—heterologous (from pollution, such as naphthoquinone), iatrogenic (such as laxative anthraquinones) or homologous (from catabolism, such as the superoxide radical, hydrogen peroxide, and lactic acid)—whenever they accumulate in the tissues, cause blockage of the electron transport chain, the consequence of which is the capture of molecular oxygen and relative hypoxia. All the pathological sequelae derive from this.
- The action mechanism of the homeopathized quinone is, as usual, interpretable in the light of the laws of enzyme kinetics.
- If a natural quinone is used (i.e., a physiological participant in mitochondrial respiration, such as ubiquinone), the homeopathized form accelerates the rate of enzymatic synthesis of the homologous quinone. The result is the unblocking of oxidative phosphorylation and the resumption of energy production.
- If a synthetic quinone (e.g., parabenzoquinone) is used, its homeopathized form functions by stimulus via the synthesis of antitoxin-specific enzymes that neutralize, by a digestive combustion process, the synthetic homologous quinones that have accumulated in the respiratory chain and impede its physiological action. The neosynthesized enzymes depolymerize the impregnating quinones.
- With homeopathic quinones, we can combat the accumulation of toxic valences (free radicals) to which humans are exposed by the pollution of the environment and food; their action is essentially definable as antioxidant.

### *Glyoxal: Poison and Drug*

Glyoxal (O(H)C–C(H)O) is used as a chemical intermediate in drug, detergent, and disinfectant production and as a binding agent in the production of a large number of different polymers. It is released during these procedures into both the air and the water of the environment: 46% is released into the hydrosphere and 54% is released into the soil (<1% is released into the air). In European rivers and aquifers, we find more than 12 µg/L of glyoxal; also, glyoxal is a by-product of disinfection with ozone and is found in drinking water. Regarding microbial activity, nonenzymatic auto-oxidation of oils, and the process of browning linked to the oxidation of sugars, glyoxal is frequently found in fermented foods and in beverages. It has been found in beer, wine, and black tea in concentrations from 20 µg/L (black tea) to 1556 µg/L (sherry). In addition, it has been found in various other fermented products such as soya paste and yogurt (0.63–4.2 mg/kg), bakery products such as bread (0.07–1.6 mg/kg), various vegetable foods (3–14 mg/kg), and food oils (>6.5 mg/kg). Glyoxal released into the environment is rapidly converted by abiotic processes of bacteria and fungi and does not itself have a marked tendency to bioaccumulation.

Glyoxal intoxication can result from inhalation or transdermal absorption with the use of disinfectants (e.g., in hospitals) or antiparasitics in the environment (e.g., in agriculture). In addition, the general population is exposed to this intoxicant through foods, air, and polluted water. Thus, throughout the world, large amounts of glyoxal are produced for industrial use, but the pollution from this substance derives from many sources. Car emissions form a photochemical smog that produces glyoxal, which is also found in cigarette smoke. Domestic fires also release glyoxal. We can see, therefore, that with the usual diet of today, made up of coffee, toast, yogurt, oil, and bread, and with "normal" life in our towns and our houses, we will inevitably absorb a toxic dose of glyoxal. This aldehyde is also produced during normal cell metabolism by a multitude of non-enzyme–dependent reactions, such as spontaneous reactions of amino groups with reducing sugars (Maillard reaction), auto-oxidation of sugars, oxidation of DNA, oxidation of polyunsaturated fatty acids, UV photodamage, oxidative stress conditions, and reduced glutathione depletion. In addition, glyoxal is a product of microsomal metabolism and oxidation of products such as glycol aldehyde, ethylene glycol, and β-hydroxy–substituted *N*-nitrosamines. The concentration in human blood is from 0.1 to 1 μmol/L, with higher levels in subjects suffering from diabetes mellitus or renal insufficiency. In biological materials, less than 10% of glyoxal is present in free form in aqueous solution; most of the carbonyl groups are reversibly bound to protein residues of cysteine, lysine, and arginine types.

Glyoxal attacks the amino groups of proteins, nucleotides, and lipids and is considered an important intermediate in the formation of advanced glycation end products. These substances alter protein function and inactivate the enzymes, consequently altering the entire cell metabolism and inhibiting protein synthesis and cell proliferation. The damage linked to advanced glycation end products (see **Fig. 4.51**) increases with the length of life of proteins and, therefore, is particularly harmful to collagen, neurofilaments, and crystalline structures. Shorter-lived proteins, such as hemoglobin, lipoproteins, and intracellular proteins, are damaged as well. These products have a role in the progression of many chronic pathological conditions, including diabetes, Alzheimer disease and other neurodegenerative pathologies, chronic inflammatory pathologies, arthritis, arteriosclerosis and vascular pathologies in general, cataracts, senile skin changes, pulmonary fibrosis, renal insufficiency, and complications of peritoneal dialysis.

The deleterious effects of these aldehydes in the cell are counteracted by the reduced glutathione-dependent glyoxalase enzyme system, which con-

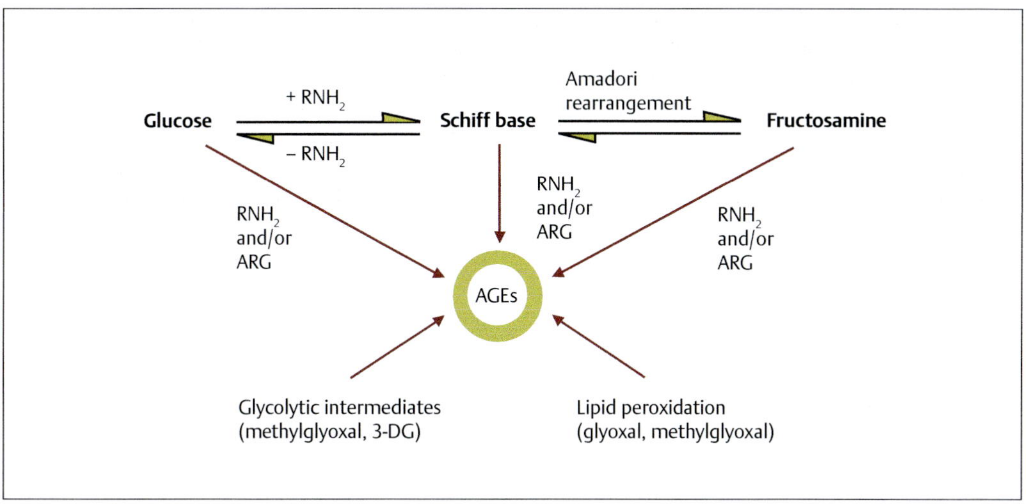


**Fig. 4.51** Early and advanced glycation.

verts glyoxal to the much less reactive glycolate. Therefore, endogenous concentrations of glyoxal in human body tissues and fluids, like those of other α-oxoaldehydes, are limited by the efficiency of the glyoxalase system and the speedy reaction of glyoxal with proteins.

Glyoxal also reacts rapidly under physiological conditions with arginine to form 1-(4-amino-4-carboxybutyl)-2-imino-5-oxo-imidazolidine. Thus, arginine residues at the protein level or the exogenous administration of arginine can act as detoxifiers of this aldehyde. In the presence of further glyoxal, phenylalanine or lysine can also be oxidized and can produce amides.

In patients with chronic renal insufficiency, the increase in glyoxal probably arises because of the known acceleration of auto-oxidation of glucose as well. Hyperproduction and local accumulation of glyoxal has been noted in diabetes mellitus. In experimental cardiac ischemia in the pig, a very considerable increase in glyoxal in the myocardium has been noted, which progressively increases to reach a peak of 24 times normal 6 hours after the event.

The acute toxicity of glyoxal in experimental animals is low to moderate and therefore depends on the concentration. After inhalation, it causes irritation and local hyperemia in the eyes, larynx, and respiratory organs. After oral ingestion, there is irritation and congestion of the gastrointestinal tract, the lungs, the kidneys, and the adrenal glands. In the liver, pancreas, and kidney in particular, glyoxal causes serious degenerative actions similar to those caused by diabetes mellitus. Glyoxal when absorbed through the skin also causes a marked increase in blood sugar. Glyoxal in vitro is directly genotoxic to bacterial and human cells and causes mutations, chromosome damage, and DNA changes. The substance has been shown to have a direct carcinogenic action only in the stomach; this action is not seen in the liver and skin despite repeated local applications. The incubation of rat retina cells considerably stimulated cell apoptosis. This event is linked to the induction of large amounts of advanced glycation end products (**Fig. 4.51**) and confirms the neurotoxicity of these products presumably responsible for the retina damage so common in diabetics. From the homotoxicological viewpoint, these tests are very important because the administration of infinitesimal homeopathic doses of aldehyde is designed to regulate the decisive balance between proliferating cell stimulus and apoptotic stimulus, which is often absent in neoplastic subjects.

The target organs of the toxicity of this aldehyde are the pancreas and the kidney. In the pancreas, necrotic areas are seen, particularly at the beta cells of the islets of Langerhans, with damage similar to what can be induced with Alloxan and similar to diabetes. The nephrotoxic action is distinguished by vacuolar degeneration of the kidney. Other toxicological studies on animals have shown similar degenerative effects at the hepatic level. Retarded growth and weight gain, due to the systemic toxic effects of the aldehyde (not reduced consumption of food), have also been noted in experimental animals. In intoxicated animals, a reactive increase in the glyoxalase purification system has been noted, but only for a limited period of time. These observations would seem to confirm the utility of the homotoxicological practice of administering diluted aldehyde to almost all patients, at approximately monthly intervals, in order to activate the detoxifying processes, given the continuous state of reintoxication in which modern humans live. The common laboratory tests do not show particular variations, except for a decrease in blood protein levels that is linked to the inhibition of protein synthesis induced by the aldehyde. Despite the widely documented genotoxicity of glyoxal, no notable carcinogenicity was observed, even during the course of prolonged administration of aldehyde. This apparently odd result can be explained in the light of what follows.

A specific study on glyoxal was published in 2004 in Geneva, Switzerland, under the auspices of the World Health Organization by J. Kielhorn, C. Pohlenz-Michel, S. Schmidt, and I. Mangelsdorf of the Fraunhofer Institute for Toxicology and Experimental Medicine in Hanover, Germany. This study was performed in order to assess the toxicity of glyoxal in the environment and human health. Despite all the data that emerged, which we have partly mentioned, there are, as usual in this field, no definitive data on human toxicity. However, animal studies and epidemiological data from our population speak fairly clearly about the toxicity of these chemical substances whenever they are, for reasons of pollution, at excessive levels in the body. Levels are excessive because, as shall be seen later, endogenous formation (see **Fig. 4.52**), more often than not, has a reactive, compensatory, and defensive effect on the cell, causing its apoptosis to ward off cancer.

### The Process of Glyoxalation in the Cell

In many bacteria, fungi, and plants, there is a simplified catabolism pathway, providing an alternative to the Krebs cycle (**Fig. 4.53**). This pathway proceeds directly from isocitrate to malate. The reason this pathway has developed in some living forms is to be found in the need to have energy in the absence of glucose, by using fatty acids and amino acids instead.

The administration of glyoxal in small doses presumably has the effect of stimulating this energy supply pathway in humans too, an extremely useful method in diabetic patients, in whom there is little possibility of using glucose because of the scarcity of insulin. In patients with chronic disease also, there can be a supplementary energy production pathway in the absence of the various catalytic pathways of the Krebs cycle.

Conventional studies are compelled to find drugs to block this metabolic pathway in order to counter very aggressive fungi and bacteria. Homotoxicology tries to stimulate the induction of this metabolic capacity to overcome shortages linked to toxic or degenerative problems.

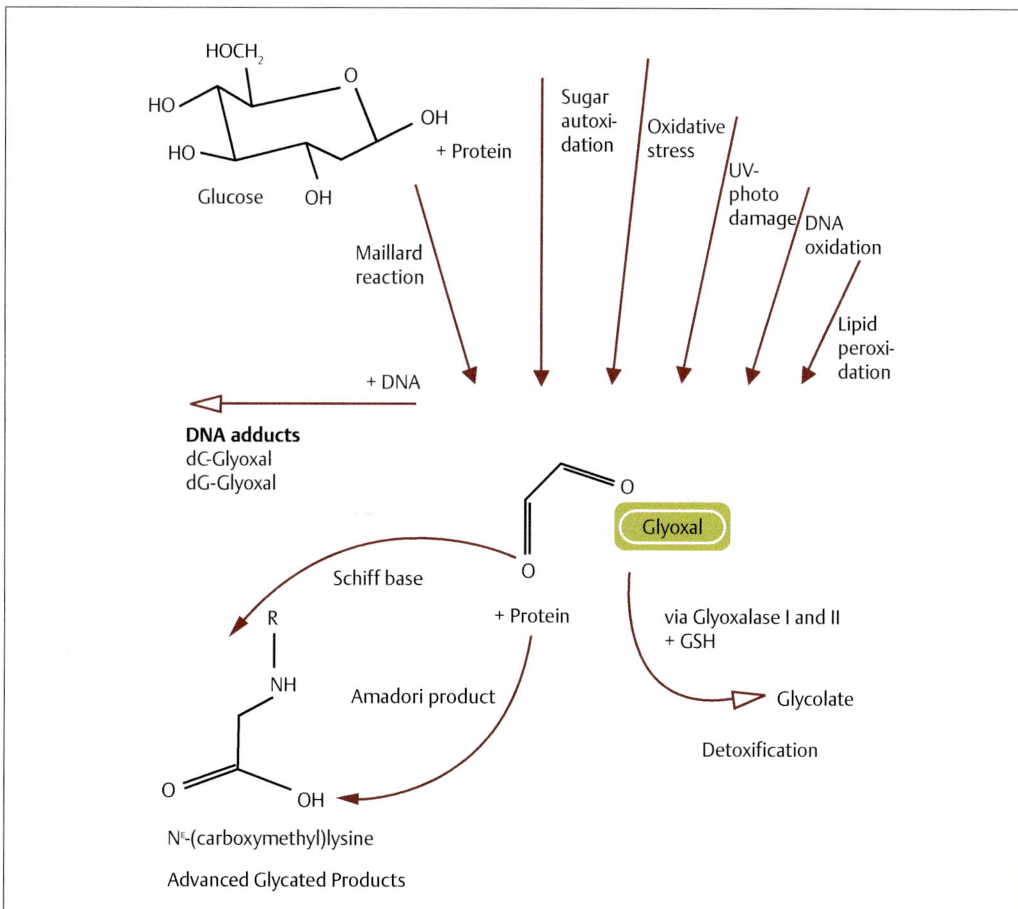


**Fig. 4.52** Endogenous sources of glyoxal and its metabolism.

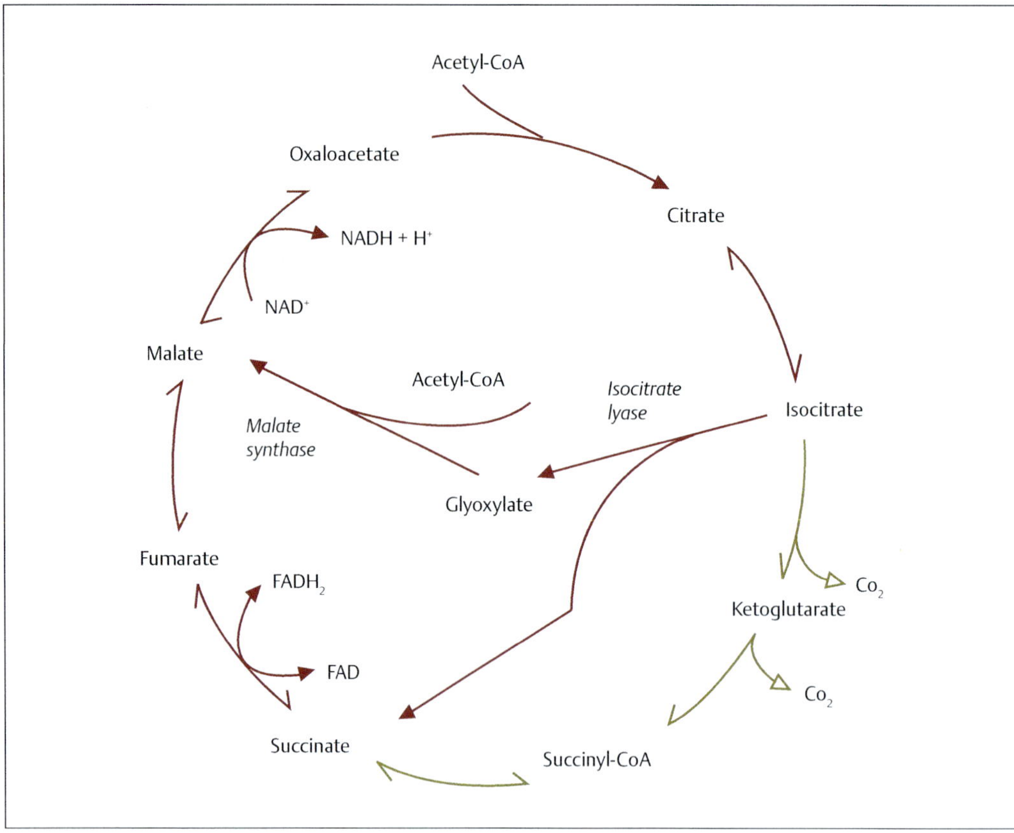


**Fig. 4.53** Alternative catabolic pathway to the Krebs cycle.

### *Aldehydes*

Aldehydes, which, along with ketones, are also called carbonyl compounds because of the presence of a carbonyl group, are involved in numerous chemical reactions, where R and R′ in **Fig. 5.54** can be aliphatic or aromatic.

Electrons of the C=O double bond, uniting two atoms of very different electronegativity, are concentrated on the more electronegative atom, oxygen. In effect, the carbonyl group is extremely polarized: it has been calculated that the positive charge on the carbon atom is about 47% of a electronic charge. The oxygen can therefore act as a good acceptor to form bonds with hydrogen.

Aldehydes form Schiff bases on reacting with amines (**Fig. 4.55**). These organic compounds act on the tissue as powerful free radicals. The administration of aldehydes in small amounts, as suggested by homotoxicology, is a stimulus to dissociation and so to the emunctory elimination of these dangerous tissue toxins.

Aldehydes are thus very reactive chemical forms; the indirect synthesis of these substances by phagocytes can play an important role in the elimination of bacteria and neoplastic cells. Glycolaldehyde (O=(H)C–COH) inactivates glucose-phosphate dehydrogenase, glyceraldehyde-3-phosphate dehydrogenase, and copper-zinc superoxide dismutase in mammary carcinoma cell cultures, thus suppressing the growth of the cells and inducing their apoptosis.

Test results suggest that glycolaldehyde can be an important mediator of the antineoplastic activity of neutrophils. According to studies by Manju Ray, an Indian scientist at the University of Calcutta, the use of small amounts of methylglyoxal in combination with ascorbic acid can block the

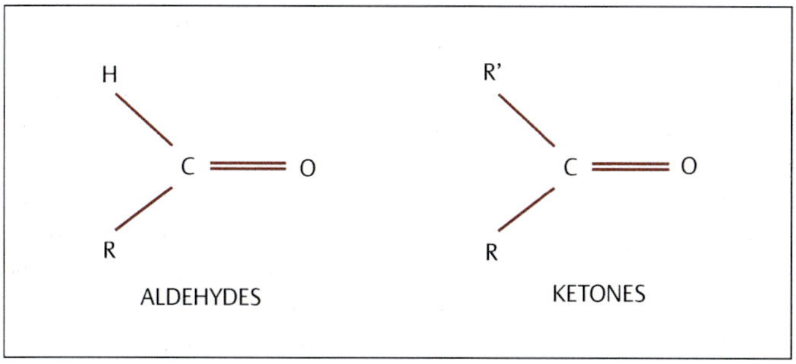


**Fig. 4.54** Carbonyl compounds: aldehydes and ketones.

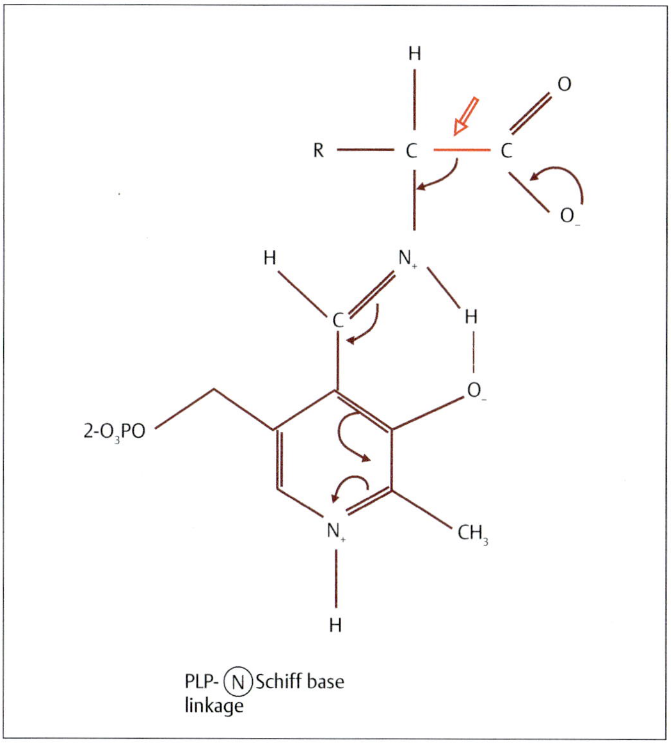


**Fig. 4.55** Schiff bases.

glycolysis of neoplastic cells without affecting healthy ones.

On exposure of leukemia L1210 cells to 0.1 µm of methylglyoxal-bis(guanylhydrazone) in vitro, there is a dose-dependent inhibition of cell proliferation accompanied by structural changes in the mitochondria, which lose their cristae and show less matrix density. This damage precedes growth inhibition by 12 hours.

Glyoxal attacks the protein, nucleotide, and lipid amino groups and is considered an important intermediate in the formation of advanced glycation end products (see **Fig. 4.51**). The modification of advanced glycation end products alters protein function, inactivates enzymes, and, in the final analysis, disturbs the cell metabolism, blocks proteolysis and protein synthesis, and inhibits cell proliferation.

The deleterious effects of the highly reactive glyoxal are counteracted by the ubiquitous reduced glutathione-dependent glyoxalase system, which converts the glyoxal into the less reactive glycolate. The glucose reacts initially with an amino group with a lysine side-chain to form a Schiff's base that, through an Amadori reaction, forms fructosamine. Methylglyoxal is an α-oxo-aldehyde, a powerful glycating agent. Triosephosphates are the largest sources of methylglyoxal. The formation and concentration of methylglyoxal are greater in those with hyperglycemia.

Maitake, a fungus used in Japanese medicine and scientifically known as *Grifola frondosa*, has antiproliferative properties extensively tested by numerous scientific studies. The action of this mycotherapeutic agent is explained by a cytotoxic activity mediated by the inhibition of glyoxalase I, a vital disintoxicating enzyme for the cell.

The blocking of this enzyme leads to the accumulation of methylglyoxal, which has a cytotoxic and blocking effect on mitosis, in particular of the neoplastic cells that typically use the anaerobic pathway to produce energy and that largely depend on the use of glucose. As previously mentioned, the accumulation of these metabolites forces use for energy purposes of fats in particular.

Glyoxals are, thus, aldehydes that form at the endogenous level and increase in situations of hyperglycemia or thiamine deficiency. These substances are powerful factors in oxidative stress, particularly of the hepatocyte, which can more easily be subjected to the consumption of glutathione by hydrogen peroxide and can also interfere with the antioxidant protection processes of the cell. By supplementing thiamine, the hepatocyte is protected from the damage caused by excessive production of glyoxal.

Recent studies have demonstrated that activated phagocytes use myeloperoxidase to produce glycolaldehyde, 2-hydroxypropanal, and acrolein. Because α-hydroxy- and α,β-unsaturated aldehydes are very reactive substances, it can be presumed that the synthesis of these substances mediated by the phagocytes would play an important role in destroying bacteria and tumor cells. Some scientists in Kuwait, using mammary carcinoma cells, have demonstrated that glycolaldehyde inactivates glucose-6-phosphate dehydrogenase, glyceraldehyde-3-phosphate dehydrogenase, and copper-zinc superoxide dismutase, suppressing cell growth and inducing apoptosis. These studies suggest that glycolaldehydes can be important mediators of the antitumor activity of the neutrophils.

The glyoxalase enzyme system is a metabolic pathway that is found in the cytosol of all cells. Recent research has seen the development of new antitumor and antimalarial drugs directed at this enzyme. Blood glyoxalase levels are considered risk factors for the development of diabetic complications and there is a renewed interest of the pharmaceutical industry in this enzyme system.

Glyoxalase is of vital importance for sustaining the first stages of embryogenetic development and subsequently for maturation and development, up to the processes of aging and death. This enzyme system is, therefore, involved in the control of cell growth, the detoxification of the α-oxoaldehydes, and the bypassing of the triosephosphate → pyruvate section of the Embden–Meyerhof pathway[75].

This system is, in fact, relatively simple and comprises two consecutive enzyme reactions (see **Fig. 4.56**) that lead to the end production of D-lactate, using reduced glutathione as a cofactor. The importance of these reactions is associated with the metabolism of the physiological substrate methylglyoxal, a reactive α,β-dicarbonyl compound that is inevitably formed in the biological glycolysis system. If the glyoxalase system is altered or blocked, methylglyoxal accumulates at toxic levels, a situation that occurs during antineoplastic and antimicrobial therapies. In dose-limiting conditions, methylglyoxal induces initially reversible, then irreversible, protein changes.

### Albert Szent-Gyorgyi and Hans-Heinrich Reckeweg

Szent-Gyorgyi, who was awarded the Nobel Prize in 1937 for the discovery of vitamin C, also performed exciting studies in the field of oncology in the mid-twentieth century. In particular, he identified the mechanisms that put the cell at rest or stimulate its reproduction, even if uncontrolled. He applied the name *Retine* to substances that, chemically none other than glyoxal and methylgloxal, put the cell at rest. He also isolated methylgloxal from urine and tested its antineoplastic

[75] Bianchi I, Pommier L. *Grande Dizionario enciclopedico di omeopatia e bioterapia*. Palermo: Nuova Ipsa Editore; 2007

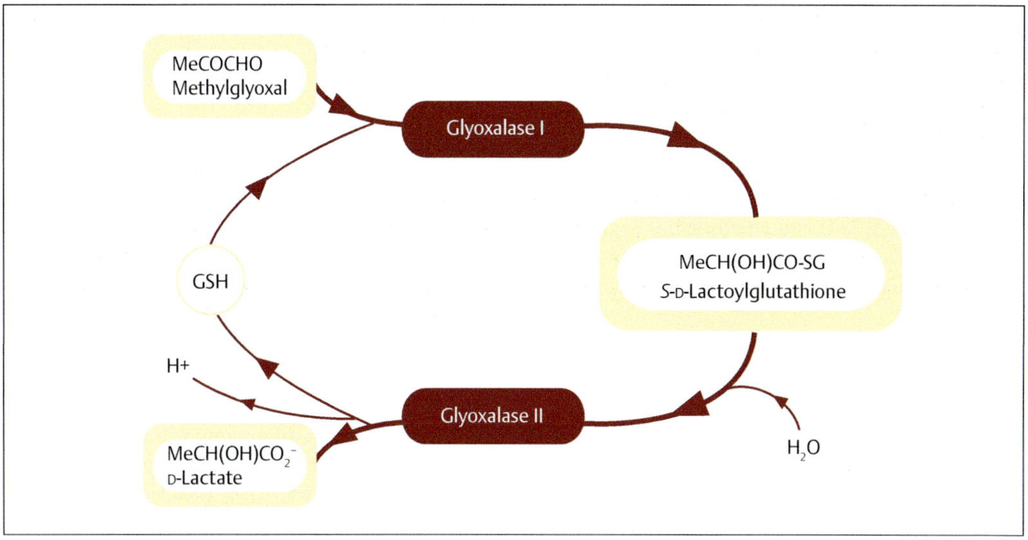


**Fig. 4.56** The glyoxalase system.

properties. He also referred to *Promine*, which designates substances (i.e., glyoxalase) that stimulate the chaotic proliferation of the cell.

In the journal *Science*, he wrote the following: "Retine and Promine are two physiological substances, produced in nature, which are probably able to influence cell division; we may be in the presence of substances that have the ability to stop tumor growth and also to induce the return to normal of neoplastic cells, without toxic effects." In his research, Szent-Gyorgyi demonstrated that the glyoxal and methylglyoxal carbonyl groups halt cell division and cause a state of quiescence in the cell. In the absence of carbonyl groups in the tissue, an anarchic and uncontrolled cell proliferation is stimulated, the end result of which can be the onset of neoplasia. In 1972, in *Prevention* magazine, Szent-Gyorgyi stated that "Glyoxal and Methylglyoxal are substances produced by the cell, which are not toxic and have specific antineoplastic activity." Subsequent research also attributed antiviral activity to these substances. Further research by Egyud (National Academy of Science, 1965), a colleague of Szent-Gyorgi, demonstrated that glyoxal and methylglyoxal normally exist in the cell, thus supporting the indication for their use in homeopathic infinitesimal amounts, to stimulate their activity according to the Arndt–Schultz rule. In 1968, Szent-Gyorgyi further explained the oncostatic action of methylglyoxal by demonstrating experimentally that neoplastic cells in culture are more sensitive than normal cells to this substance. Subsequent tests demonstrated that mice that were inoculated intraperitoneally with sarcoma 180 were cured by intraperitoneal injections of methylglyoxal.

In fact, methylglyoxal is simply an α-ketoaldehyde (pyruvic aldehyde) and all cells have a powerful enzyme system to convert it into the corresponding nonreactive oxyacids such as lactic acid. This enzyme system is called glyoxalase. In fact, there are two enzyme systems (glyoxalase I and II) and one coenzyme (GSH, reduced glutathione), serving to transform glyoxal derivatives into the corresponding hydroxyacids and methylglyoxal into lactic acid. According to Szent-Gyorgyi, the difference between a healthy and a neoplastic cell is that the latter has lost the ability to bind the previously mentioned enzyme complex, which, therefore, inactivates the methylglyoxal. In addition to inducing the characteristic tissue acidosis that follows cell cancerization, an uncontrolled, unnecessary, or even neoplastic growth of the cell is induced. Szent-Gyorgyi also mentions the research of Warburg and his anaerobiosis theory, according to which the true cause of cancer is the ability of the animal cell to live without oxygen by using anaerobic glycolysis.

In an interview given to *Prevention* magazine in 1973, Szent-Gyorgyi said that he was very close to understanding the cell mechanisms of cancerization and their control, but complained about the minimal financial support provided for his research.

Szent-Gyorgyi stresses the importance of the carbonyl group C=O and the dicarbonyl group O=C–C=O. Koch also emphasized the importance of carbonyl groups in controlling neoplastic proliferation.

> ***RETINE*** → *Glyoxal and methylglyoxal* → ***Cell proliferation blocking***
> ***PROMINE*** → *Glyoxalase enzyme complex* → *Degradation of methylglyoxal to lactic acid* → *Stimulation of anaerobic glycolysis* → ***Uncontrolled cell proliferation***

Warburg, Koch, and Szent-Gyorgyi were thinking along the same lines. Reckeweg, also aware of the scientific and legal fate of such great scientists, did not make claims, but introduced drugs such as quinones and aldehydes that embrace these theories into therapy.

What can we conclude after all this careful examination and what can the practical effects be from the therapeutic viewpoint? The first clear notion is that nature has organized a fairly complex regulation system to control cell proliferation. The therapist must intervene with extreme caution. The cell must reproduce because it is the only bastion against senile degeneration of the tissues and their continual proliferation. The primary cause of the aging of the medulla and nerves is the absence of a continuous and efficient supply of stem cells. At the same time, if this reproduction is not controlled, cell differentiation toward the embryonic phase is induced, and the risk of cancer would be very high. This explains why Reckeweg, after studying Szent-Gyorgyi's theories, does not suggest a drug with concentrated or barely diluted glyoxal or methylglyoxal[76]. This type of drug could, in fact, be a too abrupt brake, useful for the neoplastic patient but dangerous, for example, for an elderly subject, whose tissues are already tending to poor exchange and slow proliferation. In addition, thinking homeopathically, we do not want to block a mechanism with very high dilutions of these substances. This is why Reckeweg suggests a drug that is very important in its impact on the cell, Glyoxal compositum. This drug contains glyoxal and methylglyoxal in the regulatory D10 dilution. Glyoxal compositum is an important drug for detoxification of the accumulation of glyoxal, which can originate from external pollution and from internal metabolism, particularly in the elderly subject. Accumulation occurs all the more in those with diabetes or renal insufficiency. However, rather than suppressing glyoxal, it is more a matter of damming up the accumulation of these aldehydes that, in small amounts, are useful obstacles to abnormal cell proliferation; in large amounts, they have a toxic and degenerative effect on the cells and tissues.

The practical deductions from all this are that Glyoxal compositum, in an aging, dysmetabolic, and intoxicated subject, is useful in spaced administration, one intramuscular vial every 1–3 weeks. In the neoplastic subject, it is useful in massive administration, with possible recourse to D6 dilutions or using not one but at least five vials, in the same injection, once or twice a week. In this way, we activate internal chemotherapy and obstruct cell proliferation. In a very thorough manner, conversion to anaerobic metabolism and acidosis (prerequisites for the growth and diffusion of neoplasia) are blocked.

The clinical indications of glyoxal and methylglyoxal in homotoxicological form include: chronic renal insufficiency, diabetes mellitus, diabetic complications, hepatic toxicosis, pollution damage, smoking damage, disinfectant intoxication, anticryptogam intoxication, Alzheimer disease, neurodegenerative pathologies in general, chronic inflammatory pathologies, arthritis, arteriosclerosis, vascular pathology in general, cataracts, elderly skin degeneration, and pulmonary fibrosis.

### *Conclusion and Recapitulation*

The degradation of macromolecules with a nutritive function (polysaccharides, lipids, and proteins) progresses in stages. In the first stage, the polymers are degraded to monomers; in the second

---

[76] Együud LG, Szent-Györgyi A. Cancerostatic action of methylglyoxal. *Science.* June 7 1968;160(3832):1140

stage, the monomers undergo transformation to acetyl-coenzyme A; and in the third stage, $CO_2$ and $FADH_2$ are produced by the Krebs cycle. The reoxidation of NADH leads to the formation of ATP and $H_2O$. In the Krebs cycle, the eight enzyme reactions oxidize the acetyl-coenzyme A, which derives from pyruvate, to $CO_2$ and $H_2O$. Once formed, NADH and $FADH_2$ produce ATP by oxidation. The respiratory chain is located in the mitochondrial inner membrane and is the main oxidative means for the formation of ATP.

The sequence of redox systems in the respiratory chain is logically established according to their redox potentials. The most negative potential is that of the NADH + $H^+$/$NAD^+$ system. The sequence follows an $H^+$ and electron transport system that terminates in a quinone: coenzyme Q. The critical point of these processes is actually the quinone, where protein transport gives way to electron transport. Ubiquinone acts as a metabolic toxin if present in free form at a high concentration. In these cases, it is necessary to bind the radical differently to restore cell respiration. This is possible with the administration of another quinone in homeopathic dilution. For example, supplying parabenzoquinone or triquinoyl transforms ubiquinone into hydroquinone, with release of the respiratory chain and restoration of the immune function. In the neoplastic cell, the concentration of lactate increases, with relative deficit of glyoxal and methylglyoxal; by administering these two homeopathized substances, the metabolism is returned to the right direction.

The carbonyl groups act on the division of the host–pathogen integration. As is well known in certain conditions of allergy, or neoplasia, a bond is created between toxin, or virus, and cell respiratory group, through a double azomethine bond. This is able to produce neoplasias or any allergy whatsoever. The administration of activated carbonyl groups (i. e., quinones in homotoxicological form) induces separation of the toxin or virus and the host cell, restoring physiological function. In practice, there is dehydrogenation of the hydrogen atom and the resulting carbon free radical binds molecular oxygen, becoming a peroxide free radical. The subsequent separation creates nitrogen oxide and $CO_2$, restores the carbonyl group, and destroys the pathogen. Thus, we may finally deduce the following therapeutic indications, which are also applicable in the pediatric area:

- Acute and chronic viral diseases
- Allergic and autoimmune diseases
- Neuromuscular diseases
- Dermatological diseases
- Chronic degenerative diseases

## Organ Support

Historically the cell and its environment have been seen as two separate entities, and it is only in recent years that the importance of cell-to-cell communication as well as the importance of the interaction between the matrix and the cell has become apparent.

It is worthwhile at this point to undertake a comprehensive analysis of this part of the history, because it is now clear from the concept of the extracellular matrix first put forward by Pischinger in the 1980 s[77] up to the concept of living matrix proposed by Oschman[78] that it provides us with the foundations to understand the holistic functioning of the body and the terrain of action of most therapeutic approaches in biological medicine.

The marked interest in the study of the cell and its structures led to lack of interest in the study of the connective tissue, its structure, and its fundamental importance for homeodynamic (formerly, homeostatic) functioning and maintenance of the cell itself.

The terrain concept described by Claude Bernard has been the foundation of the extracellular matrix point of view. This concept was initially formulated in 1859 and fully developed by 1865. It is important to underscore the key role of this French researcher in what would later be the develop-

[77] Pischinger A. *Matrix and Matrix Regulation – Basis for a Holistic Theory in Medicine.* Brussels: Haug International; 1991

[78] Oschman JL. The nuclear, cytoskeletal and extracellular matrices: a continuous communication network. Center for Advanced Studies in the Space Life Sciences Workshop on "The Cytoskeleton: Mechanical, Physical, & Biological Interactions", November 15–17, 1996, Marine Biological Laboratory, Woods Hole, MA. [Poster Session]. http://www.mbl.edu/CASSLS/Oschman.poster.html

ment of medical physiology as a science[79-82]. According to Bernard, the terrain has clearly protective functions and allows bodily functions to be regulated in order to preserve vital activity. In his view, the perturbations of these functions first cause changes in the terrain fluids (ECM) in which the cells are embedded and finally lead to illness and death. Bernard emphasizes that internal regulation is based on the terrain's capacity for full communication and its ability to adapt and adjust to changes in the environment. Bernard first identified the inner terrain with blood plasma, but later broadened his interest to include lymph and finally, all body fluids. The concept of terrain is related to the concept of interior milieu and is the initial contribution of physiology to biophysics, exerting great influence on the development of holistic medicine.

In 1927 Hericourt published his book on the terrain, in which he emphasizes how the same pathological factor often produces an extremely wide variation in the responses of different individuals[83]. In 1936 the French biochemist Wladislaw Kopaczwewski wrote a paper on how, in a single subject, the terrain counterbalanced the pathogenic influences, whether external or internal, acting as a protective barrier against different noxious agents[84].

Between 1870 and 1920, the work of Pasteur and Koch on microorganisms led to an overvaluation of these external pathogenic factors and underestimated the role of internal factors. Pasteur, in his scientific counterargument to Bernard, acknowledged the role of terrain as a passive, static, unchanged factor, comparable to a laboratory culture medium.

### *Homeodynamics of the Matrix Fluids*

Several discoveries and concepts of researchers on seemingly unrelated topics contribute to the understanding of homeodynamics of the matrix fluids. Because of his terrain concepts, Bernard became the father of regulation physiology and opened the path for the birth of humoral physiology based on the introduction of physical chemistry.

The principles of ionic solutions and dilution were developed by Jacobus van't Hoff and Svante Arrhenius. The concept of thermoregulation of the internal environment was studied by pioneers such as Rene Quinton, Joseph Barcroft, Ernest Starling, Lawrence Henderson, and Walter B. Cannon.

Starling, along with W. M. Bayliss, isolated enzyme secretions in 1902 and in 1905 introduced the notion of hormones. He was the pioneer in studies of peripheral circulation, understanding the role of osmotic forces and their importance in the exchange between capillaries, intercellular tissue, and the lymph system.

Henderson, influenced by the Hippocratic idea of vital force, studied acid–base regulation at Harvard University, in the fatigue study laboratory, and published his book Fitness of the Environment in 1913. Henderson understood that the basis of life lies in equilibrium of the extracellular environment and its components, stressing the prominent role of water in the matrix[85].

In 1929 Cannon introduced the concept of homeostasis as a mechanism stabilizing physiological variables. He considered the fluid matrix of the body to be important and gave importance to the concepts of stress and considered that bodily reactions are always protective, purposive, appropriate and "wise"[86, 87].

---

79 Bernard C. *Leçons sur les propriétés physiologiques et les altérations pathologiques des liquides de l'organisme.* Paris: Baillière; 1859

80 Bernard C *Introduction à l'étude de la médicine expérimentale.* Paris: Baillière; 1865. [English ed.: *An Introduction to the Study of Experimental Medicine.* New York: Dover; 1957]

81 Holmes FL. Claude Bernard and the milieu intérieur. *Archives Internationales d'Histoire des Sciences.* 1963;16: 369–376

82 Olmsted EH. Historical phases in the influence of Bernard's scientific generalizations in England and America. In: Grande F, Visscher MB, eds. *Claude Bernard and Experimental Medicine.* Cambridge, MA: Schenkman; 1967:24–34

83 Héricourt J. *Le Terrain dans les Maladies.* Paris: Flammarion; 1927

84 Kopaczewski, MW. Le terrain, le microbe et l'état infectieux: Claude Bernard ou Pasteur? *Revue Scientifique* 1936;74(14):417–425

85 Henderson LJ. The Fitness of the Environment. New York: Macmillan; 1913

86 Cannon WB. *Bodily Changes in Pain, Hunger, Fear, and Rage.* 2nd ed. New York: Appleton; 1929

87 Cannon WB. *The Wisdom of the Body.* Philadelphia: WW Norton; 1932

This line of research was later continued by Hans Selye, who in 1936 studied the nonspecific, systemic response of the organism to a wide range of stress factors. He showed how many acute and chronic diseases are accompanied by a state of exhaustion and have the same group of symptoms; he also came up with the general system theory on the basis of his observations[88–91].

*Matrix Biophysics*

In keeping with Bernard's tradition, the biophysicists from the French school Fred Vlès and Wladislaw Kopaczwewski continued the humoral physiology approach with studies of acid–base balance and reduction–oxidation potentials and their influences on the proteins and "colloids" of the terrain of the body[92–100].

This school contended that water was the basis of biology, as it determines the properties of living substances; they emphasized the importance of pH measurements, as well as measurement of the redox potential of the blood and other body fluids, established by Michaelis[101, 102] and Clark[103, 104] and popularized by Louis-Claude Vincent[105, 106] whose work was reconfirmed more recently by Orszagh[107, 108], Fougerousse[109], and Eccles[110].

*A New Concept of the Cell*

An important number of concepts in the cell paradigm deserve our attention. The cell is considered in essence to be a solution surrounded by a membrane permeable to ions and other protein-free

---

88 Selye H. A syndrome produced by diverse nocuous agents. *Nature.* 1936;148, 84–85

89 Selye H. The general adaptation syndrome and the diseases of adaptation. *J Clin Endocrinol* 1946;6:117–196

90 Selye H. *Stress: The Physiology and Pathology of Exposure to Stress.* Montreal: Acta Medica; 1950

91 Weiner H. *Perturbing the Organism – The Biology of Stressful Experience.* Chicago: University of Chicago Press; 1992

92 Vlès F. *Cours de Physique Biologique.* Paris: Vigot Frères; 1927

93 Vlès F. Revue des notions actuelles sur un problème de physico-chimie pathologique: Les propriétés des points isoélectriques et du terrain physico-chimique dans l'organisme normal ou pathologique; leur application à l'étude des tumeurs. *Archives de Physique Biologique.* 1929;7(Fasc. Suppl. No. 5):1–64

94 Vlès F. Recherches sur les propriétés physico-chimiques des électrolytes. *Archives de Physique Biologique.* 1936;13(1): 57–73, 88–107

95 Reiss P. *Le pH intérieur cellulaire.* Paris: Les Presses Universitaires de France; 1926

96 Reiss P. L'action du potentiel d'oxydation-réduction du milieu sur l'autolyse protéique du cristallin et du corps vitré. *Archives de Physique Biologique.* 1940;15 (4): 39–40

97 Kopaczewski MW. La tension superficielle en biologie. Parts 1–3. *Archives de Physique Biologique.* 1921; 1(4):145–167; 1922;2(2) 1–20; 1922;2(3):1–12

98 Kopaczewski MW. *Théorie et practique des colloïdes en biologie et en médecine.* Paris: Vigot Fréres; 1923

99 Kopaczewski MW. *Introduction à l'étude des colloïdes. L'état colloïdal et ses applications.* Paris: Gauthier-Villars; 1926

100 Kopaczewski MW. *Traité de Biocolloïdologie.* 2nd ed. Paris: Gauthier-Villars. [Vol. 1: *Pratique des colloïdes* (1930). Vol. 2: *Biocolloïdes.* Vol. 3: *Phénomènes colloïdaux* (1932–33). Vol. 4: *État colloïdal en médicine* (1937–38)]

101 Michaelis L. *Die Wasserstoffionenkonzentration.* Berlin: Julius Springer; 1914

102 Michaelis L. *Oxydations- und Reduktionspotentiale.* Berlin: Julius Springer; 1933

103 Clark WM. *The Determination of Hydrogen Ions.* Baltimore: Williams & Wilkins; 1925

104 Clark WM. *Studies on Oxidation-Reduction.* Hygienic Laboratory Bulletin, No. 151. Washington, DC: United States Public Health Service; 1928

105 Vincent LC. Aperçu sur la bio-electronique. *Revue de Pathologie Générale et Comparée.* 1954;54(663):121–130

106 Vincent LC. Bio-Electronique. Définition des trois facteurs phroniques pH, rH2 et p. Bulletin de la Société de Pathologie Comparée, Séance du Mardi 14 Juin 1955. *Revue de Pathologie Générale et de Physiologie Clinique.* 1956;56(677):643–656

107 Országh J. Réactions d'oxydo-réduction et acido-basiques – vers une approche théorique et expérimentale plus cohérente. *Siences du Vivant.* 1990; (1 3éme trimestre):23–35

108 Országh J. Quelques aspects physio-chimiques des coordonnées bioélectroniques. *Sciences du Vivant.* 1992;(4 2éme trimestre):45–62

109 Fougerousse A. L'approche bio-électrique de Vincent. *Sciences du Vivant.* 1992;(4 2éme trimestre):63–79

110 Eccles R. *Electrolytes, Body Fluids and Acid Base Balance.* London: Edward Arnold; 1993

molecules, both small and large. The membrane-pump theory of the cell was first proposed by the discoverer of osmosis W. Pfeffer in 1877, and later established in 1902 by Helmholtz and refined by Donnan in 1911, Hodgkin in 1949, and finally, by A. Huxley and K. Cole.

In 1835 Felix Dujardin introduced the concept of living jelly and later Jan E. Purkinje gave it the name of protoplasm. Following the introduction of the colloid concept by Thomas Graham in 1861, many biologists and physiologists conceived of protoplasm as a hydrated polyphasic colloidal system[111–116].

In 1928 Ernst and Scheffer suggested that intracellular potassium ions were in the bound form and in the nonionized form.

In 1908 Heinrich Schade recognized the cell as a polyphasic colloid and published his pioneer book *Physical Chemistry in Internal Medicine* in 1921, where we can find that he considered the body a "microheterogeneous system, a mixture of gel-like and sol-like masses."

In 1910 Martin H. Fischer postulated that the intracellular substance in living cells is a water-binding colloid and could not be compared to a suspension or solution of molecules in water. Subsequently, he hypothesized that it is a hydrated system in which water is not free but is bound to colloids in the form of alkaline or acid proteinates with a high hydration potential[117,118].

On the basis of the work started by Schade, Fischer, and other biologists, it was understood that the water in the cell and in the extracellular spaces is *not* free[119,120]. The strong affinity of water for many molecules was discovered in 1848 by Carl Ludwig of the Berlin School of Physiology.

However, since the 1950s a significant number of other concepts have been developed and have become well-established, tracing back to the analysis of the colloid properties of living matter.

In 1951 and 1952, the Russian physiologist Afanasy S. Troshin presented his pioneering studies on cell permeability, founding the "sorption theory" of the cell[121,122] as variants of the membrane theory, based on the interaction of macromolecules and ions with cell water and the lipoproteins of the membrane. The conclusion was drawn that protoplasm coacervation gives water the very special property of "liquid crystallinity." Jakob Segal's German group continued the work started by the Russians[123,124].

The latest alternative theory was presented by the U.S. physiologist Gilbert N. Ling, known as the association-induction theory[125–129], was fully con-

---

[111] Fischer MH, Moore G. On the swelling of Fibrin. *Am J Physiol.* 1907;20:330

[112] Fischer MH. *Oedema: A Study of the Physiology and the Pathology of Water Absorption by the Living Organism.* New York: John Wiley; 1910

[113] Fischer MH, Hooker MO. *The Lyophilic Colloids (Their Theory and Practice).* Springfield, IL: Charles C. Thomas; 1933

[114] Schade H. Kolloidechemie und Balneologie. *Medizinische Klinik.* 1909;(29):1070 and (30):1116

[115] Schade H. *Die physikalische Chemie in der inneren Medizin. Die Anwendung und die Bedeutung physikochemischer Forschung in der Pathologie und Therapie.* Dresden: Theodor Steinkopff Verlag; 1921

[116] Fischer, MH, Suer WJ. *Der kolloide Aufbau der lebenden Substanz.* Darmstadt: Steinkopff; 1951

[117] Fischer MH. *Oedema: A Study of the Physiology and the Pathology of Water Absorption by the Living Organism.* New York: John Wiley; 1910

[118] Fischer, MH, Suer WJ. *Der kolloide Aufbau der lebenden Substanz.* Darmstadt: Steinkopff; 1951

[119] Newton R, Gortner RA. A method for estimating hydrophilic colloid content of expressed plant tissue fluids. *Bot. Gaz.* 1922;74:442–446

[120] Newton R. 1938

[121] Troshin AS. *Problems of Cell Permeability.* Oxford: Pergamon Press; 1966 [Russian original ed.: Moscow: Academy of Sciences, 1956]

[122] Nassonov DH. *The Local Reaction of the Protoplasm and the Spreading of Excitation* [in Russian]. Moscow: Academy of Sciences; 1959

[123] Segal J. *Die Erregbarkeit der lebenden Materie.* Jena: Gustav Fischer; 1958

[124] Segal J, Körmer U, Leiterer KP. *Doe Entstehung des Lebens aus biophysikalischer Sicht.* Jena: Gustav Fischer; 1983

[125] Ling GN. *The Interpretation of Selective Ionic Permeability and Cellular Potential in Terms. Waltham, MA: Blaisdell; 1960*

[126] Ling GN. *A Physical Theory of the Living State: The Association-Induction Hypothesis.* Waltham, MA: Blaisdell; 1962

[127] Ling GN. 1970

[128] Ling GH. *A Revolution in the Physiology of the Living Cell.* Malabar, FL: Krieger Publishing; 1992

[129] Ling GH. *Life at the Cell and Below-Cell Level. The Hidden History of a Fundamental Revolution in Biology.* New York: Pacific Press; 2001

firmed by Freeman Cope[130] using nuclear magnetic resonance imaging to observe a difference in the signals given off by the water in a normal cell compared with that in a cancer cell[131].

These studies laid the foundation for studying the properties of the physical state of water in cells and its role in the structure[132], understanding how changes in the physiological status of the organism can alter the physical properties of cellular water, extensively investigated by Cope, Drost-Hansen and Clegg, Hazlewood, and Tigyi.

The concept of a highly structured cell has been supported since the first discoveries of organelles such as the Golgi apparatus or mitochondria in the late 19th century; later with the advent of electron microscopy, in 1945 Albert Claude discovered the endoplasmic reticulum, microtubuli, microfilaments, and intermediary filaments, and finally, Porter discovered the microtrabecular lattice between 1981 and 1984[133].

It is now clear that the cytoskeleton is a highly structured three-dimensional network that is highly dynamic and labile. It can be readily disassembled under the influence of small perturbations, possibly oscillating between the gel state in which the structure becomes visible and the sol state in which it is dissolved.

One of the most revolutionary aspects, which we will return to below, is the molecular biology research described in Pienta and Coffey's works in 1991[134], showing the connection between extracellular matrix, cytomatrix, and the nucleus with its content, giving us a holistic vision of the body structure as a whole for the first time. This provides us with an understanding of how the information from the environment or system itself is completely integrated.

130 Cope FW. The solid-state physics of electron and ion transport in biology. *Adv Biol Med Phys.* 1970;13: 1–42

131 Damadian R. Tumor detection by nuclear magnetic resonance. *Science.* 1971;171(3976):1151–1153

132 Whipple HE. Forms of water in biological systems. *Ann N Y Acad Sci.* 1965;125(2):249–772

133 Porter KR, Tucker JB. The ground substance of the living cell. *Scientific American.* 1981;244:40–51

134 Pienta KJ, Coffey DS. Cellular harmonic information transfer through a tissue tensegrity-matrix system. *Med Hypotheses.* 1991;34:88–95

### *Regulative Mechanisms*

In 1918 the German pathologist Friedrich Kraus synthesized the findings of physical chemistry, the physiology of humoral and neural regulation, and colloid science in what is known as constitutional medicine. He emphasized the functional unity of regulatory mechanisms at the endocrine, vegetative-nervous and physical-chemical level[135].

F. Kraus reformulated humoral physiology, presenting the body as an oscillatory system of ionotrophic colloids whose state changes between sol and gel, based on the action of mechanical effects and the effects related to the electrical interface potentials. Kraus also felt that the neurovegetative system was of primary importance in regulating the system. Kraus's works were pioneering and served as an inspiration for Schade, Ricker, Eppinger, Hauss, and Pischinger who continued the work.

### *The Extracellular Matrix*

In the early decades of the twentieth century, some physiologists and pathologists started to pay attention to what was known as connective tissue, now usually called extracellular matrix.

One of the main pioneers in this field was the founder of molecular pathology, Heinrich Schade, who recognized the organ function of connective tissue and established the nature of the extracellular matrix as an active physiological system[136–140].

135 Kraus F. *Allgemeine und spezielle Pathologie der Person. Klinische Syzygiologie.* Leipzig: Thieme, 1919 [2nd ed. 1926]; [Vol 1: *Allgemeiner Teil.* Vol. 2: *Besonderer Teil: Tiefenperson*]

136 Schade H. Physikochemische Beiträge zur Oedemfrage. *Verhandlungen des Deutschen Kongresses für Innere Medizin (Wiesbaden).* 1912;No. 29:526–535

137 Schade H. Untersuchungen zur Organfunction des Bindegewebes. II. Mitteilung: Das Quellungsvermögen des Bindegewebes in der Mannigfaltigkeit seiner Erscheinungen. *Zeitschrift für experimentelle Pathologie und Therapie.* 1913;14(1):1–29

138 Schade H. *Die physikalische Chemie in der inneren Medizin. Die Anwendung und die Bedeutung physikochemischer Forschung in der Pathologie und Therapie.* Dresden: Theodor Steinkopff Verlag; 1921

139 Schade H. Die Physikochemie des Bindegewebes und

Schade criticized Virchow's position, which considered it to be a purely connective tissue, in which only the connecting and support function was stressed, and opposed his colloid-chemical concepts. In 1911 Schade started his physicochemical research on the matrix, summarizing his findings in *Physical Chemistry in Internal Medicine* in 1921. Schade's works laid the groundwork for the tremendous research efforts of those after him, namely, Gustav Ricker, Hans Eppinger, Heinrich Thiele, Werner H. Hauss, Alfred Pischinger, Otto Bergsmann, and Hartmut Heine, among many others.

Gustav von Bergmann, a professor of internal medicine, confirmed that functional disturbances precede pathological abnormalities; he affirmed that these functional abnormalities could take years to appear in the cell phase, drew attention to the care that should be taken with any larvate form of disease, and abolished the artificial demarcation between functional disease and organic illness[141].

Felix Buttersack continued this research, localizing latent illness in the connective tissue and introducing the notion of "ground tissue," discriminating the differences between support tissues and loose connective tissue[142].

Ricker, a pathologist and founder of neural pathology, considered neural regulation of fluids and vascular permeability to be key in physiology; he postulated his well-known relationship condition[143, 144], against Virchow's cell-centric concept, holding that cells are interdependent with the nervous system and humoral system in his phase theory.

Hans Eppinger and his permeability pathology described in 1949 also give us a neural approach that shows how the vegetative nervous system controls vascular changes[145]. Between 1940 and 1960 the colloid scientist Heinrich Thiele showed how the interaction between complex colloids and ions is the basis for the structural mechanisms in the organism structures, but particularly in the matrix[146–148].

The cardiologist Hauss worked on the biochemistry of connective tissue, particularly in what is known as the nonspecific mesenchyme reaction observed in many illnesses[149–151]; demonstrating that this is the basic reaction to any kind of stimulus and that the metabolic response of the matrix is highly sensitive to any kind of stimulus, such as infections, toxins, foreign substances, and even stimuli of emotional origin.

The Austrian histologist and embryologist Alfred Pischinger, considered to be one of the founders of histochemistry, started his work on connective tissue in 1948, developing the theory of ground regulation[152–154]. According to Pischinger's

ihre Bedeutung für die Lymph- und Ödembildung. *Verhandlungen der Deutschen Gesellschaft für Innere Medizin.* 1922; 34:283–287

140 Schade H. Die Physikochemie des Bindegewebes und ihre Bedeutung für die Lymph- und Ödembildung. *Verhandlungen der Deutschen Gesellschaft für Innere Medizin.* 1922; 34:283–287

141 Bergmann G von. *Funktionelle Pathologie.* Berlin: Julius Springer; 1932

142 Buttersack F. *Latente Erkrankungen des Grundgewebes, insbesondere der serösen Häute.* Stuttgart: Ferdinand Enke Verlag; 1912

143 Ricker G. *Entwurf einer Relationspathologie.* Jena: Gustav Fischer; 1905

144 Ricker G. *Pathologie als Naturwissenschaft. Relationspathologie.* Berlin: Julius Springer; 1924

145 Eppinger H. *Die Permeabilitäts-Pathologie als die Lehre vom Krankheitsbeginn.* Vienna: Springer Verlag; 1949

146 Thile H. Richtwirkung von Ionen auf anisotrope Kolloide – Ionotropie. *Naturwissenschaften.* 1947;34: 123

147 Thile H. Ordered coagulation and gel formation. *Discuss Faraday Soc.* 1954;18:294–301

148 Thile H. *Histolyse und Histogenese. Gewebe und iontrope Gele – Prinzip einer Strukturbildung.* Frankfurt am Main: Akademische Verlagsgeselllschaft; 1967

149 Hauss WH, Losse H. *Struktur und Stoffwechsel des Bindegewebes.* Stuttgart: Thieme; 1960

150 Hauss WH, Junge-Hulsing G. Über die universelle unspezifische Mesenchymreaktion. *Dtsch Med Wochenschr.* 1961;86:763

151 Hauss WH, Jungen-Hulsing G, Gerlach U. *Die unspezifische Mesenchymreaktion.* Stuttgart: Gustav Thieme Verlag; 1968

152 Pischinger A. *Matrix and Matrix Regulation – Basis for a Holistic Theory in Medicine.* Brussels: Haug International; 1991

153 Bischof M [coauthored by F. Rohner]. Wasser [Water]. In *Dokumentation der besonderen Therapierichtungen und natürlichen Heilweisen in Europa,* Vol. II. *Wissenschaftliche Grundlagen der besonderen*

approach, the basic regulatory system of the extracellular matrix is the most fundamental and oldest regulatory and information system in the organism, older than the hormone and nervous systems. It is the basis of the nonspecific immune response and the transit pathway for all systemic functions. It controls and regulates all bioenergetic processes, as well as the physicochemical and bioelectrical status of the organism.

The anatomical substrate of the ground regulation system is the loose connective tissue, which consists of a highly polymerized network of glycoproteins with a specific content of cells such as the fibrocytes, mast cells, and macrophages, and a cell with great totipotentiality, the fibroblast. It includes structural glycoproteins such as collagen, elastin, fibronectin, and laminin.

Carl B. Reichter in 1845 and Georg von Rindfleisch in 1869 first recognized that the nervous system and the capillaries were not in contact with cells; later in 1949, Eppinger showed this by electron microscopy and finally Hauss proved this fully. This gives the ECM a determining role in the transport pathway to and from the cell.

Later works of Hartmut Heine revealed the entire functional network of interleukins, prostaglandins, growth factors, interferons, proteases, and antiproteases that allows fine control of basal regulation in this extremely important system, basic to an understanding of how many of the therapeutic methods of biological medicine work.

### *The Cell: History of a Paradigm*

Between 1663 and 1667, the English scientist Robert Hooke reported his observations to the Royal Society of London, commenting that cork had a structure composed of honeycomb-like elements; we now know that Hooke was observing the cell walls of dead tissue which, for that reason, contained nothing inside. This was the start of one of the most fascinating stories in the scientific progress of biology: knowledge of the cell, its structure, and its function[155,156].

The discovery made by Hooke, who documented his observations with highly precise drawings, did not elicit any major comments or interest from naturalists at the time, although the search for the smallest life structure continued. The idea of cells (Hooke) coexisted simultaneously with those of fibers (Haller) and vesicles or utricles (Malpighi).

Around the late eighteenth and early nineteenth centuries, renewed interest in resolving the mysteries of nature was evident. This was particularly true of Germany, where a philosophical current known as *Naturphilosophie* (or philosophy of nature) arose and had a strong impact on the entire European intelligentsia. The advocates of *Naturphilosophie* proposed the development of a philosophy based on the teachings of nature and so vigorously encouraged research in the various branches of the natural sciences. These included microscopic studies[157].

It was not until the mid-nineteenth century, however, that this concept would become part of standard scientific knowledge, since two fundamental aspects not yet theoretically resolved were still missing as they had not been backed by observations: first, the generalization of the existence of cells to explain the organization of all living things; and, second, the origin of these cells. This was when the names of Matthias Schleiden (1804–1881) and Theodor Schwann (1810–1882) came into the picture.

In 1838 Schleiden, taking Robert Brown's discovery of the cell nucleus as a reference, began describing the nucleus and proposing a function for it. The perseverance and precision of his observations were such that he was able to identify the nucleolus within the nucleus. In Schleiden's opinion, the basic issue was that the explanation of a plant's genesis and development should be "reduced to cell theory."

*Therapierichtungen und natürlich Heilweisen.* Essen: Verlag für Ganzheitsmedizin VGM; 1992:91–148

154 Heine H. *Lehrbuch der biologischen Medizin.* 2nd revised ed. Stuttgart: Hippokrates; 1997

155 López Pinero JM. *La medicina en la Historia.* Madrid: Ed. La esfera de los libros; 2002

156 Albarracin Teulon A. *Le teoría celular, Historia de un paradigma.* Madrid: Alianza; 1983

157 Albarracin Teulon A. *Le teoría celular, Historia de un paradigma.* Madrid: Alianza; 1983

The other figure in this story is the German zoologist Schwann, a distinguished student of the famous naturalist from Berlin, Johannes Müller (1801–1858). Starting from an outstanding series of microscopic observations, of which he offered an in-depth interpretation in *Microscopic Researches into the Accordance in the Structure and Growth of Plants and Animals* (1839), Schwann extended the cell theory described by the botanist Schleiden from plants to animal organisms: both plants and animals are composed of cells; cells have a life of their own to a certain extent, and this individual cell life is subordinated to that of the body as a whole. In particular, his book *Microscopic Researches* clearly described the similarity in development and structure between plants and animals. He was a pioneer in cytology and histology.

Around 1845, the German pathologist and researcher Rudolf Virchow (1821–1902), who was convinced that cells are at the core of all life activity and based his opinions on observations by his colleague Remak, reached the conclusion that cells originate only from preexisting cells. This conclusion has been stated by Virchow in Latin and become a famous maxim: "*omnis cellula e cellula*" (every cell comes from another cell). It is important to note that, despite these contributions, toward the end of his life Virchow challenged the idea that a life force actually exists. He proposed that the phenomenon of life is so complex that no mechanical explanation could give a full account thereof and thus it would be advisable to accept that life constitutes a phenomenon that is something "special," something that could never be fully explained from physical and chemical studies "even should it be possible to comprehend life as a whole in terms of a mechanical result of known molecular forces"[158, 159].

Interest in studying the cell has intensified from the time it was first considered a primary unit of life. Improvements in optical instrumentation and staining techniques allowed rapid advances in the observations and descriptions of both the eukaryotic cell nucleus and the cytoplasm. The organelles were discovered one after another, revealing a cytoplasmic complexity far greater than the simplicity ascribed by the first cytologists, who described protoplasmic mass as homogeneous. It was still unknown how cell division takes place.

In the early twentieth century the rediscovery of the works of Gregor Mendel (1822–1884) and the cumulative knowledge of the cell opened up a new field of biological exploration: cytogenetics[160, 161].

Walther Flemming (1843–1905) focused his efforts on studying the cell nucleus and gave the name "chromatin" to the substance occupying its interior, because of the material's tendency to fix certain stains and so become differentiated from the rest of the cell content. But Flemming's fundamental contribution was to describe mitosis and identify chromosomes[162, 163].

The idea of mutation proposed by Hugo De Vries (1848–1935) and recorded in the works of Thomas Morgan (1866–1945) on the *Drosophila* fly to explain changes in organisms made it possible to "merge" within a single general explanatory framework not only cell theory but also Mendelian genetics and Darwin's theory of evolution[164, 165].

Among all the techniques used to analyze DNA during those years, crystallography was the first to provide an indication of its structure. In the 1940s, an analysis of purified protein crystals suggested to the U.S. physicist Linus Pauling and to the Englishman Maurice Wilkins that this molecule had a helical thread shape[166–169].

---

158 Albarracin Teulon A. *Le teoría celular, Historia de un paradigma*. Madrid: Alianza; 1983

159 Brown TM, Fee E. Rudolf Carl Virchow, medical scientist, social reformer, role model. *Am J Public Health*. 2006 Dec; 96(12):2104–2105

160 López Pinero JM. *La medicina en la Historia*. Madrid: Ed. La esfera de los libros; 2002

161 Albarracin Teulon A. *Le teoría celular, Historia de un paradigma*. Madrid: Alianza; 1983

162 Albarracin Teulon A. *Le teoría celular, Historia de un paradigma*. Madrid: Alianza; 1983

163 Paweletz N. Walther Flemming: pioneer of mitosis research. *Nat Rev Mol Cell Biol*. 2001 Jan;2(1):72–75

164 De Vries H. The principles of the theory of mutation. *Science*. 1914 Jul 17;40(1020):77–84

165 De Vries H. The origin by mutation of the endemic plants of Ceylon. *Science*. 1916 Jun 2; 43(1118): 785–787

166 Pauling L, Delbrück M. The nature of the intermolecular forces operative in biological processes. *Science*. 1940 Jul 26; 92(2378):77–79

Around 1944 the U.S. biochemist Oswald T. Avery, while researching the infectious action of pneumococci, discovered that DNA was the material that carried the hereditary traits of all living beings, without exception as it was also found in the mysterious viruses. With this discovery, biochemical studies on the chemical makeup and structure of DNA came to the forefront[170].

Erwin Schrödinger, an important German physicist who emigrated to the United States during World War II, expressed vividly the key role given at that time to DNA: chromosome fibers "contain, in some kind of code-script, the entire pattern of the individual's future development and of its functioning. Chromosome structures also have the means to execute this program. They are law-code and executive power ... architect's plan and builder's craft in one ..."[171,172].

The crystallographers' work did not go unnoticed by the researchers James Watson and Francis Crick, who elaborated on that work to suggest that the DNA molecule was indeed helical. Finally, in April 1953 they proposed the definitive model for the DNA molecule—the double helix model—and a few weeks later, suggested how it was replicated.

However, there was still no interpretation of how the information contained in DNA flowed and allowed this molecule to fulfill the functions of replicating and translating itself into proteins. As a result of various investigations motivated by the double-helix model, in 1957 Crick first enunciated the "central dogma of molecular biology" based on the key concepts of replication, transcription, and translation[173].

In 1961 the French researchers F. Jacob, J. Monod and A. Lwoff postulated the key role of messenger RNA. Four years later they finished the complete deciphering of the genetic code. Later, with the advent of DNA sequencing techniques, molecular genetics entered its decisive phase of development that led it far beyond theoretical interest and ushered in a number of techniques that would transform this knowledge into one of the keys for the development of biomedicine and industry[174].

### *Practical Organ Regulation*

There is a long history of therapeutic use of organ extracts in homeopathy. In homotoxicology, the work of August Bier is thought to have influenced Reckeweg to include them in the antihomotoxic repertoire. In his work on treating chronic disease, Bier had noticed that organ extracts are more effective than secretory products of the same organs and concluded that treating disease in a specific organ requires a prepared extract of that organ. Through their work in the Charité clinic in Berlin, Bier and Reckeweg were able to demonstrate empirically that organ therapy is a superior form of treatment for chronic disease.

Cellular dysfunction will result in organ dysfunction or, even worse, necrosis of the affected tissue with degenerative pathologies as a consequence. To understand the need for cell and organ support we have first to understand the healthy cell.

---

167 Pauling L. *Icosahedral symmetry. Science.* 1988 Feb 26;239(4843):963

168 Pauling L. Icosahedral and decagonal quasicrystals of intermetallic compounds are multiple twins of cubic or orthorhombic crystals composed of very large atomic complexes with icosahedral point-group symmetry in cubic close packing or body-centered packing: Structure of decagonal Al (6) Pd. *Proc Natl Acad Sci U S A.* 1989 Dec;86(24):9637–9641

169 Arnott S, Kibble TW, Shallice T. Maurice Hugh Frederick Wilkins CBE: 15 December 1916 – 5 October 2004. *Biogr Mem Fellows R Soc.* 2006;52: 455–478

170 Avery OT, Dubos R. The specific action of a bacterial enzyme on pneumococci of type III. *Science.* 1930 Aug 8;72(1858):151–152

171 Ganten D. What is life? On Erwin Schrödinger, his cat, and the Journal of Molecular Medicine. *J Mol Med.* 2007 Dec; 85(12):1291–1292

172 Tesche C. QUANTUM MECHANICS: Enhanced: Schrodinger's Cat Is Out of the Hat. *Science.* 2000 Oct 27; 290(5492):720–721

173 Watson JD, Crick FH. Molecular structure of nucleic acids; a structure for deoxyribose nucleic acid. *Nature.* 1953;171(4356):737–738

174 Jacob F, Perrin D, Sánchez C, Monod J, Edelstein S. [The operon: a group of genes with expression coordinated by an operator. C.R.Acad. Sci. Paris 250 (1960) 1727–1729]. *C R Biol.* 2005 Jun; 328(6):514–520

A living organism is more than the sum of the separate cells of which it is composed. Nevertheless, the cell remains the functional building block of the tissue and its correct functioning is essential for the life of the whole organism.

After fertilization, during the early phases of rapid cell division, certain cells begin to migrate to the interior of the blastula, giving rise to the three germ layers: endoderm, mesoderm, and ectoderm. This stage of embryogenesis is called the "gastrula stage." Thus: the zyygote gives rise to the morula (a solid ball of 12–32 blastomere cells), which gives rise to the blastula (over 100 cells), which gives rise to the gastrula. It is during the gastrulation phase the germ layers begin to form.

The **ectoderm** will give rise to:
1. Skin and its appendages
2. Autonomic nervous system
3. Central nerve system

The **endoderm** will give rise to:
1. Gastrointestinal tract
2. Respiratory system
3. Endocrine glands

The **mesoderm** will give rise to:
1. Bones and muscles
2. Urogenital system
3. Mesenchyme (connective tissues)
4. Cardiovascular system

In bioregulatory medicine, this scheme the basis of the vertical axis on the DET (see Chapter 4).

Although extremely complex in their function, even the "bigger" human cells are small. The red blood cell belongs to the smaller cells in the human body and is about 6–8 µm across. Cells of tissues and organs represent a synergic cooperation of billions of cells, each of them organized as an independent unit but simultaneously in continuous contact with *all* the other cells of the body. The amount of information exchanged is enormous, not only due to the number of cells, but also due to the complexity and variety of information sent and received.

Cells may be embedded on a basal membrane as in tissue structures, for example, as seen in an organ structure. Substance exchanges are mainly achieved through diffusion over the extracellular matrix.

Cells may also occur not embedded on a basal membrane. They may be part of a global tissue structure (e.g., the fibroblast) or may be completely free in movement (e.g., blood cells).

Both embedded and nonembedded cells will be surrounded by a matrix, carrying information from one cell or system to another cell or system. This matrix also protects the cell from any toxic burden by acting as a biophysical filter.

We should keep in mind that the extracellular matrix is:
- The direct and immediate environment of every cell in the organism
- The transmission area for most steering messengers and interactions between body systems
- The site for autoregulation and deregulation in the body
- The site where inflammation occurs
- The deposition site for many toxins

Thus, the purity of the extracellular matrix is essential for the life quality of the cell.

*Ways in Which the ECM May Influence Cellular Health*
- Physical blockages at the level of the ECM, obstructing the transmission of nutrients from the bloodstream through the matrix to the cell and of metabolic products in the other direction, that is, from the cells to the matrix and from there to the venous system and/or lymphatic system.
- Lack of sufficient oxygen, causing hypoxia and malfunction of the mitochondria (resulting in less energy production).
- Blockage of enzymatic processes, obstructing correct information transmission for the steering of various other metabolic and functional processes.
- Direct cell damage due to trauma of diverse origin.
- Damage from free radicals generated during inflammatory processes or from free radicals from other processes that oxidize the cell structures directly.
- Intracellular storage of homotoxins such as heavy metals or even microorganisms such as viruses using the cell as a host.

### Cell Death

In cell death we should clearly differentiate between two phenomena that have completely different characters: **apoptosis** and **necrosis**.

- **Apoptosis** is a *normal* process of cell death, mostly meant to replace cells in a multicellular environment. Apoptosis prevents old and dysfunctional cells remaining and disturbing the functioning of younger, healthy cells, or preventing immune cells becoming activated against self-antigens. It is a preprogrammed cell death and is part of the natural processes and cycles of cellular life.
- **Necrosis** is a premature, *pathological* cell death, due to many causes and agents damaging or destroying the cell. Necrosis is one of the main characteristics of degenerative pathologies.

Injury or variations in time and intensity of exposure to toxins can lead a normally functioning cell directly to necrosis. Less intensive or shorter exposures may lead to adaptive changes in the cell, to degenerative characteristics, and also to necrosis if prolonged. Necrosis is an irreversible condition (cell death) that can occur as a result of acute, intensive destructive agents (injury, anoxia, ...) or in the long term as a result of repetitive destructive insults.

Cell death is in a healthy adult organism mainly compensated by cell division. Conversely, uncontrolled proliferation of cells must be checked by cell death. This regulation phenomenon, in which the healthy situation must be in balance between cell division and cell death, is called homeostasis. Apoptosis occurs to maintain homeostasis. Necrosis may be a disturbing factor for homeostasis, as it is an unplanned, pathological cell death.

### Treating Disease at the Organ Level

In treating disease at the organ level, three levels are treated as a holistic unit:

1. **The interstitial level.** In histology this is the so called extracellular space (ECM), which is the direct environment of the living cell. It is important is to understand that the vital quality of the cell is depends directly on the purity or the functional state of this interstitial environment. Problems at this level will induce cell dysfunction and through this subsequent organ dysfunction.
2. **The cellular level** is about the condition of the cell itself, the functioning of the cell and its interaction with other cells over the message transmission system operated through the ECM or interstitial system. Living cells create waste products that a well-functioning interstitial system has to eliminate. All the cells together form an integrated homogenous system called an organ. The functioning of the organ is directly related to the condition of cells.
3. **The organ level** is about the synergy of the cells. This synergy is formed by the interactions it has with other organs and tissues of the same holistic system. Within the psycho-neuro-endocrine-immunological (PNEI) system, every organ, even every cell, is influenced by emotions and thoughts and vice versa: organ dysfunction will influence the emotional status of the living being (think about the influence of clinical symptoms on the emotions of the patient; pain, fever, nausea, ...).

Of course, an improvement in cell function will result in better organ function, but with organ support we try to improve the functioning of the organ as a "unit." We can use medications simply to improve the function of the cell and the organ, or we can use medications which are also postulated to also strengthen the tissue structure. Both types are classified into a group of medications called the organoregulators.

For the purpose of special compositum preparations are developed and there is a specific formula for almost every organ. This is of course also patient specific and we will look for the *locus minoris resistentia*, the weakest organ or organ system in the body, as it is precisely this organ that normally will cause the most problems, especially in the aging process.

*Examples of the Most Important Organ Regulators and Their Targets*

- Mucosa comp
  Mucosa
- Hepar comp
  Liver
- Solidago comp
  Urinary tract
- Tonsilla comp
  Lymph system
- Thyreoidea comp
  Thyroid function
- Cerebrum comp
  Cerebral functions
- Cutis comp
  Skin
- Pulsatilla / Echinacea comp
  Defense system
- Ovarium comp
  Female hormonal balance
- Testis comp
  Male hormonal balance
- Placenta comp
  Circulation

The Disease Evolution Table classifies the progression of disease to an organ level. In simple terms, the regulation/compensation division (RCD) separates phases in which the patient is still able to regulate toward health from phases in which only compensation is possible.

To the left of the RCD, we are dealing primarily with disturbances in function; to the right, disturbances in both function and structure are present. This distinction is less clear in the impregnation phase (where cell death may be visible only at the microscopic level) than in the degeneration and dedifferentiation phases (where structural disturbances generally become macroscopic).

Clearly, therefore, diseases that involve only functional disturbances will require a different approach from diseases involving disturbances in structure as well as function. During the excretion, inflammation, and deposition phases, regulation still takes place via excretory mechanisms, so physiological support will reactivate the organism's self-regulation. However, once the condition has progressed to the right of the RCD, resolution cannot be achieved without structural support for specific organs.

Evidence is emerging to support the idea of "resonance" between organ extracts and their corresponding organs. Today, fetal human liver cells and liver extracts are being used for liver regeneration. If therapy with extracts of a particular tissue has regenerative effects on that tissue, the same should hold true of the organotropic suis-organ preparations in antihomotoxic medicines.

Composita and suis-organ Injeels are examples of medications containing organ extracts. The importance of Reckeweg's formulations, however, lies in the fact that he recognized the interdependence among different organ systems, as mentioned above. Thus the formula of Cutis compositum includes liver and adrenal extracts, recognizing the relationships between the skin and stress hormones and between the P450 detoxification enzymes in the skin and the liver. The inclusion of liver and adrenal extracts also counteracts iatrogenic damage from products used in skin disease.

To the right of the regulation/compensation division (RCD), use of catalysts as well as composita becomes mandatory. Because no cell can function without energy, the role of catalysts in maintaining cell health is increasingly being recognized, along with their therapeutic possibilities. For this reason, advanced organ support protocols include catalysts in addition to a number of composita containing suis-organ extracts.

### Classification of Antihomotoxic Medicines: Organotropic vs. Functiotropic

*Organotropic* medications act on a specific organic system or tissue, whereas *functiotropic* medications have biochemical and metabolic effects on several organ systems. Among antihomotoxic medicines, the action of the homaccords and many of the basic combination medications is functiotropic, whereas the composita and catalysts act organotropically.

Each pillar, therefore, presents both functiotropic and organotropic therapeutic options. Taking the detoxification and drainage pillar as our first example, the Detox-Kit contains two homaccords—Berberis-Homaccord, with functiotropic effects on the kidneys, and Nux vomica-Homaccord, with functiotropic effects on the intestines and liver—along with one basic combination (Lymphomyosot, which acts on the lungs, liver, matrix, and kidneys).

A homaccord is a combination of two or three single-ingredient potency chords with identical potencies. As with most antihomoxic medicines, the action of homaccords is both synergistic and complementary.

The homaccords are an essential part of antihomotoxic treatment. In the Detox-Kit, they effectively deal with mild to moderate toxicity in all diseases to the left of the RCD. In more serious illnesses, the homaccords (like the basic combinations) serve to restart regulation and are generally administered after a period of organ support.

Basic combinations are the other component of functiotropic treatment. Here, too, the combination of synergistic and complementary effects induces physiological regulation. Traumeel and Lymphomyosot are two of the best-known combinations.

Another well-known functional treatment is Hepeel.

Functiotropic and organotropic medications are administered either singly or in combination, depending on the severity of the disease process.

The organoregulators are truly regulatory and therefore are often given as an injection or oral ampoule at a frequency of 1–3 times per week.

As stressed before, it is mandatory to combine these organ regulators with cellular support in the form of catalysts.

# 5 Bioregulatory Medicine and Homotoxicology in Practice

## Introduction

This chapter contains an account of a number of individual experiences and particularly tried and tested examples of treatment, and examples are presented of how antihomotoxic medicine can be integrated into other methods of biological medicine. Indications of preparations and special forms of administration that do not always correspond to the official information given on the individual package leaflets are also presented. On the basis of years of our own practical experience and a constant exchange of information among colleagues, we recommend these therapies with a clear conscience but can accept no liability for their efficacy and safety. These individual therapies in complementary medicine must be selected carefully and used by experienced and responsible therapists.

## Names of Preparations

Preparations in antihomotoxic medicine sometimes have different names in different countries, and the homeopathic composition can also vary slightly. An account is given in the following chapters of our own experience with the preparations available in Germany. For the sake of simplicity, the suffixed letters (N, S, SN, ...) used in Germany have been omitted. The antihomotoxic medications available in other countries can be used instead of the German formulations in the specified recommended treatments without any problems.

## Availability of Medications

The regulatory environment in Germany and other countries is sometimes unpredictable, particularly for homeopathics. Unfortunately, this may mean that preparations have to be withdrawn from the market. However, the medications for antihomotoxic therapy are so plentiful and widely usable that a suitable replacement preparation can usually be found. But this does mean that the tried and tested combination therapies mentioned in this chapter may need to be updated from time to time.

## Selection of Medications

### Dosage Forms

The preparations in antihomotoxic therapy are manufactured in various dosage forms. In addition to tablets and drops there are also ointments, suppositories, nose drops, and eye drops. The choice of medication is often crucial to the success of therapy. This chapter is devoted mainly to the use of ampule preparations and their combination with other methods in biological medicine.

### *Case History*

Homotoxicology is regulatory medicine; this means that patients are first considered on a physical level and their symptoms are the focus of therapeutic considerations. Diseases caused by the spirit and soul are of course also taken just as seriously and treated if necessary.

### *Treatment Planning*

First of all, though, cleansing and detoxification therapy is the main emphasis of the homotoxicologist's treatment. The strengthening of weakened organ or tissue systems and immunostimulation or immunomodulation are also initial measures in treatment. Existing psychological problems are naturally also included in treatment plans; treatment at the constitutional level, where necessary, is given later.

There are a few differences or additions in deference to classical homeopathy that are important to homotoxicologists. These are essentially consideration of the causes and development of disease and their manifestation in the various organ systems or tissues. Hahnemann in his day observed vicariation phenomena, which were systematized by Reckeweg in his six-phase table, called today the Disease Evolution Table, allowing patterns to be identified and treatment to be carefully planned.

The first manifestation of a disease and the accompanying circumstances at the time of manifestation are just as important to the homotoxicologist as past treatments. In particular, after repeated drug therapies that block the symptoms and the autoregulatory system, the homotoxicologist observes in patients the development of chronic diseases with regulatory disturbances or blockades.

Consequently, the choice of medications is not only based on the main symptoms but also includes aspects of excretion, detoxification, organ strengthening, and immune regulation. This often means that a single medication is not sufficient in this complex assessment, particularly in chronic diseases.

Homotoxicology is in practice used successfully in acute everyday diseases such as colds. Another main focus of antihomotoxic therapy is the prevention of diseases. Homotoxicological concepts are also used successfully in chronic diseases, and in these cases they are often combined with other methods from biological medicine.

### *Dosage*

Although the dosage of the various forms of the preparations is shown on the package leaflet, this must be selected according to the patient's individual situation and on the basis of the therapist's own experience.

In general, the following standard doses of preparations in antihomotoxic medicine can be recommended. The efficacy of 10 drops corresponds to that of a tablet, that of the ampule to that of about 30 drops or 3 tablets. With a few exceptions, the combination preparations in homotoxicology generally have a standard dosage of 10 drops or 1 tablet 3× daily; in the acute phase of the disease this dose is often given every quarter or half an hour until the patient's condition improves, up to a maximum of 2–3 hours. The ampules are usually given 1–3× a week; more frequent doses of up to several times daily are possible and are recommended in acute conditions.

The dose of the appropriate homeopathic potency depends on various factors. First of all it depends on the starting material used to make the medication. These materials are of plant or animal origin, or they are minerals. The substances used can be harmless or very toxic, and they are obviously used in different dosages and potencies. Homotoxicological medications of plant origin include plants that are also used in phytotherapy, usually in low potencies, starting from the mother tincture and usually up to 6× or D6.

These potencies are used in indications very similar to those used in phytotherapy. These are sometimes called "phytohomeopathics." The same is true of the choice of dose and potency with minerals and materials of animal origin as starting materials. Such low-potency homeopathics of plant and mineral origin are used in the short term for acute diseases (Traumeel) or in the long term treatment of chronic diseases at organ level

(Cralonin). They are preparations for the homotoxicological stimulation principle.

In regulation disturbances, the homotoxicologist tends to select the medium potencies as well as homeopathics, the starting materials of which would be toxic if undiluted, such as the spotted hemlock in the preparation Vertigoheel. When using these preparations patients suffer from regulation disorders and treatment blockades and also require additional long-term therapy, although this must always be supervised by an experienced therapist.

Medications with toxic starting materials are used in acute diseases too, such as aconite or belladonna in influenzal conditions or Strychnos Nux vomica and Arsenicum album in gastrointestinal diseases. Here, however, the potencies selected are ones for which the reversal principle has turned the toxin into a healing substance.

Consequently, homeopathic medications in the medium potencies (8× to 23×, or D8 to D23) are medications for the homotoxicological regulation principle. Higher potencies are selected for the homotoxicological information principle to treat the patient on the psychological and constitutional level.

Potency chords (Injeels and Homaccords) are a special feature of homotoxicology. These medications combine the homeopathic medications in low, medium and high potencies of the starting materials for the same medication in one dosage form. With these medications, Reckeweg wanted to bring together the whole breadth and potency of the homeopathic medication. Unlike in classical homeopathy, where high potencies are given only rarely, high potencies in the potency chords do not produce any side-effects in the sense of the homeopathic drug picture, even when used several times a day and for long periods. Consequently, homaccords should often be used as acute medications with a broad spectrum of action; for example, Nux vomica-Homaccord.

In summary, however, it can generally be said that the dosage recommendation given by the manufacturer in the medication's package insert has in most cases proved accurate.

Oral formulations should not be swallowed immediately but held for a while under the tongue or in the mouth. This improves the contact time with the mucosal surface of the mouth. This applies in particular to the treatment of diseases in the head region such as sinusitis and to dizziness. Studies have shown that homeopathic medications work particularly fast and effectively if administered close to the diseased organ or tissue. This fact means that in practice a whole range of special techniques can be used to optimize the efficacy of homotoxicological preparations; these will be described below.

## Special Treatment Methods

### Ointment Dressings

Treatments with homeopathic ointments have a long tradition; their good efficacy can be made even better by carefully rubbing them in, possibly also in combination with massage. An ointment dressing has a particularly intensive effect with occlusion. After a generous application of the ointment, a plastic film is placed over the treated area and secured with a dressing. With an ointment dressing on the foot, a woolen sock is simply drawn over it. This dressing is left in place for up to 8 hours. After that time the dressing must be removed and the ointment washed off to avoid skin damage. On the small joints of the fingers, for example with Heberden's arthritis, simple plastic gloves or rubber finger stoles can be used and a normal woolen glove pulled over them.

Another tried and tested form of ointment application is acupressure. A small quantity of ointment about the size of a lentil is rubbed into the acupressure point with gentle pressure. Acupressure is particularly impressive at the acupuncture points on the ears.

### Injection Therapies

The ampule preparations of homotoxic medications can be used in a variety of ways. Most, but not all, ampules can be administered subcutaneously, intramuscularly, or intravenously. Special injection techniques further optimize the treatment.

*Autologous Blood Therapies*

Classical autologous blood therapies in which a small quantity of venous blood is collected and then immediately reinjected subcutaneously or intramuscularly can be optimized by the admixture of homotoxicological ampule preparations. The preparations Gripp-Heel, Engystol, and Euphorbium compositum have proved their worth in viral infections; with bacterial infections the patient's blood can be mixed with Echinacea compositum. Other suggestions for treatment are given in the section on specific diseases and symptoms.

**Progressive auto-sanguis therapy** according to Reckeweg is a special form of antihomotoxic autologous blood therapy. In this method, the patient's blood is potentized in various stages with the admixture of ampule preparations. There are a number of variants of this method. The following procedure has now become established in practice.

- The ampules that are to be used in treatment are opened and set up in the order in which they are to be injected. The syringe used should have a volume (about 510 mL) greater than that of the ampules used in the individual treatment stages.
- The first ampule is drawn up into the syringe. This is the symptomatic medication, that is, the one that is most suitable for treating the patient's predominant acute symptoms. With inflammations, for example, this is usually the product Traumeel.
- A vein is punctured, a little autologous blood (about 2 mL) is aspirated, and this mixture is then immediately administered intravenously. The syringe is not emptied completely; a small quantity of the mixture (about 0.05–0.1 mL) is left in it. The syringe is removed and the second ampule is immediately drawn up into this syringe through the same cannula.
- The second ampule is usually the terrain medication and is thus an excretion and detoxification medication. This is often the preparation Lymphomyosot. The syringe is filled with air, the cannula is covered, and the syringe is shaken. Ten powerful shaking strokes create the first homeopathic potentization of a nosode of the patient's own blood. The air and the resulting foam are removed and the patient is given this mixture subcutaneously or intramuscularly, intragluteally, or into the muscle of the upper arm.
- Once again, the syringe is not emptied completely and a small quantity of the mixture is left in the syringe so that this can again be mixed with a further ampule and homeopathically adjusted by means of powerful shaking strokes. In the third stage an "organotropic" is used—a combination preparation with an affinity for the diseased organ or tissue. After a further subcutaneous injection, additional stages can be added in the same way. In the fourth stage, patients often receive an immunostimulant homeopathic medication or a catalyst preparation; nosodes are also included in this last stage.

Progressive auto-sanguis therapy is used just as successfully with susceptibility to infection as in the treatment of allergic diseases. With autoimmune disease, however, it should be used very carefully and not as the first measure, so as to avoid exacerbation. This treatment activates and modulates the immune system. One or two treatments are generally given per week, with a cure comprising 5–10 treatments.

Suggested formulations for progressive auto-sanguis therapies are given in the following section on the tried and tested indications.

*Homeosiniatry*

The injection of homeopathic ampule preparations at defined acupuncture points is a tried and tested method of greatly improving the onset of action and power of homotoxicology. A homeopathic medication with a particular field of application is injected into an acupuncture point with the same effect and indication. The combination first described by Weihe, of a "pressure point" and a homeopathic medication given orally, was further developed by de la Fuye. An acupuncture needle was first dipped into the homeopathic tincture and then used for acupuncture; finally, the injection syringe and homeopathic ampule preparations were used. In recent years homeopathic single medications assigned to particular acupuncture points have increasingly been replaced by homotoxicological combination preparations. Important injection sites are alarm points, tonification and sedation points and above all Shu points on the acupuncture meridians. An intracutaneous or subcutaneous injection is first given at these points to stimulate the cutivisceral reflexes, then the needle is turned slightly to reach the deep acupuncture points. The injection volume depends on the size of the acupuncture point and ranges from 0.2 to 1 mL per acupuncture point. The admixture of local anesthetics is not beneficial in homeosiniatry because they would numb the acupuncture point and thus disrupt the influence of the homeopathic information.

#### *Infusion Therapy*

Infusion therapy is the most potent and the fastest option in antihomotoxic therapy. Infusions are particularly recommended as initial therapy in chronic inflammation with disorders of the matrix and microcirculation. Ampule preparations in antihomotoxic therapy can be combined with one another and are also compatible with all standard infusion solutions. The admixture of vitamins, minerals, plant medicines, and other ampule preparations with biological medicines, such as procaine and sodium hydrogencarbonate, is possible without reservation.

### Biopuncture

> ***Biopuncture:*** *Expanding the scope of bioregulatory medicine by using local injections.*

#### *Definition*

Biopuncture is a bioregulatory therapeutic method in which specific biotherapeutic agents are injected into tissue-related bodily zones or points on the basis of clinical and functional diagnosis[1,2]. The therapeutic strategy of biopuncture is concerned with two key factors: what to inject and where to inject it. Both factors are individualized in each patient and adapted during each session. The physician can make an enormous variety of combinations, using different products on several spots[3,4].

When biotherapeutic products are administered as local injections, they acquire an extra dimension comparable to a turbo effect. These mechanisms related to local injection are extra effects that cannot be achieved by giving the same remedies orally or by injecting them in the buttock or arm. Daily practice has proved this again and again, and more and more physicians are realizing that these local injections are expanding the scope of their practice[5,6].

#### *Location of Injection*

Biopuncture describes the use of biotherapeutic agents for injection into *specific* zones or points[7]. Antihomotoxic medications can be injected into reflex zones, trigger points, pain points, and various structures (e.g., muscles, joints, and ligaments). The specific location and depth of the injection gives the product the extra dimension that we are looking for.

#### *Frequency of Injection*

The injections are usually administered on a weekly basis. Sometimes, more than one injection is administered in a week when dealing with an acute sports injury or severe pain. When dealing

---

1 Kersschot J. *Biopuncture. A Clinical Guide.* Aartselaar: Inspiration Publishing; 2008

2 Smit A, Van Brandt B, Engelbert C, Barkauskas D. Biopuncture. SAB Meeting, Baden-Baden, February, 2008

3 Barkauskas D. *A Biotherapeutic Approach to Common Sports Injuries. Biomedical Therapy.* 2007;1(1):12–13

4 Schultz M. Biopuncture. HAS Congress, South Africa, June 8, 2008

5 Naidoo C. Biopuncture in general practice. Johannesburg, April, 2005

6 Barkauskas D. Biopuncture in family practice. HAS Congress South Africa, June 8, 2008

7 Kersschot J. *Biopuncture. A Clinical Guide.* Aartselaar: Inspiration Publishing; 2008

with chronic situations, weekly injections are also started; however, later these injections can be given every 2 to 4 weeks.

*Multiple Microdose Injections*

The major site of interaction is when the product gets into the extracellular matrix. Part of the clinical effect of biopuncture is related to the medication being taken up by an antigen-presenting cell (stimulating the immune system[8]) and through the needle stimulating nerve endings. Thus, greater results can be achieved by using several injection sites. The number of injections in one area or zone may vary from approximately from 1 to 10.

*Introduction of Biopuncture Worldwide*

The injection of biotherapeutic agents was introduced in Germany more than 50 years ago. Presently, millions of ampules are injected worldwide every year. However, the use of biotherapeutic agents did not acquire a specific name in practice until 1991, when the term ***biopuncture*** was introduced during a biomedical congress in Belgium[9]. Having a specific name to describe this technique gives it more exposure, both for medical professionals and for patients[10, 11]. Books, lectures, courses (in Belgium, the United Kingdom, Mexico, South Africa, and the United States), and publications[12, 13] on biopuncture promote the use of injections of natural products into specific sites or areas. Since 2004, injections into acupuncture points has no longer been considered as biopuncture; they are instead described as homeosiniatry[14]. Biopuncture is very accessible for conventional physicians because it does not demand any knowledge of classic homeopathy or Chinese medicine and "biopuncture" is also a description that is easy to remember for both doctors and patients. To help stimulate the use of antihomotoxic injections, a pamphlet is available to explain the benefits of biopuncture to patients (see www.biopuncture.info).

*Medications Used for Biopuncture*

Traumeel, Lymphomyosot, Spascupreel, Coenzyme compositum, and Zeel are the most popular agents in biopuncture. Traumeel and Lymphomyosot are used for the regulation of local inflammatory response. Spascupreel is injected for muscular spasm and muscular pain (smooth and striated musculature); Zeel is used for chronic joint pain and osteoarthritis with cartilage alterations (degenerative joint disease); and Coenzyme compositum is used for chronic inflammations and tissue repair (damage on the cellular level). Other agents are used by physicians who have more experience in antihomotoxic medicine.

*Different Phases of Local Injections*

In phase 1, it is sometimes necessary to "clean" the area (the local matrix) by inducing regional lymphatic drainage using local subcutaneous injections of Lymphomyosot. It is a good idea to prepare the area first, especially when dealing with "hyperresponders." In phase 2, Traumeel is used for local immunomodulation[15, 16]. In phase 3, more specific products (symptom-related medica-

[8] Porozov S, Cahalon L, Weiser M, Branski D, Lider O, Oberbaum M. Inhibition of IL-1β and TNF-alpha Secretion from resting and activated human immunocytes by the homeopathic medication Traumeel® S. *Clin Dev Immunol.* 2004,11(2):143–149

[9] Kersschot J. Biopunctuur. Biomedical Therapy Congress, Oostende (Belgium), November 17, 1991

[10] O'Byrne A. Seminaroi Taller De Biopuntura, San Salvador, El Salvador, Febuary 2008

[11] Schultz M. Biopuncture. HAS Congress South Africa, June 9, 2008

[12] Kersschot J. The complementary use of homeopathic preparations in primary care practice. *Biomed Ther.* 1997;15(2):47–52

[13] Kersschot J. Biopunktur. *Biologische Medizin.* 1997; 26(5):228–231

[14] Kersschot J. *Biopuncture in General Practice.* Aartselaar: Inspiration; 2004

[15] Van Brandt B. *Inflammation Means Healing.* 3rd ed. Aartselaar: Inspiration Publishing; 2004

[16] Kersschot J. *Biopuncture. A Clinical Guide.* Aartselaar: Inspiration Publishing; 2008

tions such as Spascupreel, Hormeel, and Engystol) or a homaccord (e.g., Nux vomica-Homaccord, Ignatia-Homaccord, and Drosera-Homaccord) can be used. In phase 4, a deeper level is involved (the cellular level or even organ support); the medications injected include Zeel, Coenzyme compositum, and other more specific "composite" medications (e.g., Echinacea compositum, Ovarium compositum, Testis compositum, and Discus compositum).

When we want to speed up the treatment, we can skip a phase or combine several phases, depending on the individual situation.

### Needles and Syringes

Safety measures and techniques intended to avoid complications should be followed strictly. Doctors are advised to perform biopuncture according to the most current regulations in medicine. The use of sterile and disposable materials is obligatory. A sterile technique is crucial. Biopuncturists always follow the guidelines of conventional medicine regarding injection techniques.

Syringes of different sizes (2, 3, and 5 mL) are used. The needle must be attached very firmly to the syringe in order to prevent the needle from coming off during injection; a screw lock on the syringe may prevent this. We use small needles to make the procedure less painful.

***Needles Used in Biopuncture***

| | |
|---|---|
| *0.3 × 13 mm* | *30G × ½ inch* |
| *0.4 × 20 mm* | *27G × ¾ inch* |
| *0.4 × 40 mm* | *27G × 1½ inch* |
| *0.5 × 40 mm* | *25G × 1½ inch* |

### Clinical Effects of Biopuncture

Most physicians use biopuncture for minor orthopedic problems. They start to use it for neck and back pain. However, acute sports injuries such as an ankle sprain can also be treated with biopuncture. Biopuncture is also very successful in treating tennis elbow, golf elbow, and muscle strain. However, biopuncture is not just used for pain or injuries of the musculoskeletal system; it can also be used to treat functional disorders. These patients may show several symptoms (e.g., pain in the gastric area and tension in the neck or lower back muscles) without any relevant radiological or laboratory signs. This is where a lot of discomfort generally originates (e.g., mild gastritis, whiplash, and lower back pain). For example, allergies and all kinds of inflammatory conditions can be treated. many family physicians confirm that patients with asthma, eczema, hay fever, bronchitis, cystitis, and sinusitis can be treated with biotherapeutic injections. Other conditions can be treated if a practitioner is experienced in the technique. These conditions include migraine[17] whiplash, tension headache, Crohn disease, colitis, vertigo, cluster headache, and neuralgia. Each patient has to be assessed individually.

### Injection Techniques

In biopuncture there are two major pathways of treatment: (1) local treatment and (2) systemic treatment. Local treatment focuses on the actual complaints of the patient and deals with them by giving local injections. The systemic approach deals with the general status of the patient and is used to bring the "terrain" of the patient back into balance, and is thus a true bioregulatory treatment[18,19].

For the local approach, strategies should be adapted to the type of patient being seen. If the patient is a "hyperreactor," we use a soft approach, which consists mainly of subcutaneous injections of, for example, Lymphomyosot, the use of small needles, and the injection of a small volume of agent. When dealing with a "hyporeactor," we provide a more intensive approach, with in-depth injections, higher volumes, and other products. The majority of the patients are normal reactors and undergo a standard approach.

A standard approach starts with a mixture of Lymphomyosot and Traumeel (phase 1 plus phase 2). Then, more specific products (symptom-related medications) can be used; these include Spascu-

[17] Kersschot J. *Biopuncture in General Practice*. Aartselaar: Inspiration; 2004:84

[18] Schultz, M. An introduction to biopuncture. HAS Congress South Africa, Durban, June 9, 2008

[19] Kersschot J. *Biopuncture. A Clinical Guide*. Aartselaar: Inspiration Publishing; 2008

preel, Hepeel, Hormeel, and a Homaccord ampule (phase 3 products). Later, the cellular level is treated, with injections of products such as Zeel, Discus compositum, and Coenzyme compositum (phase 4 products). In some cases, a phase can be skipped or more than two phases can be combined, depending on the individual situation.

A clinical examination and description of symptoms can guide the doctor in determining the correct injection technique with regard to the choice of the localization and depth of injection. Subcutaneous injections are easy and safe, and are an interesting choice for inexperienced practitioners or when the patient is afraid of deep injections. Subcutaneous injections can also be used for the preparation of the local area (before giving deeper injections) or when deeper injections are impossible (too painful or too dangerous). When muscular dysfunction is suspected (e.g., muscular pain and loss of power in an area), injections into the affected muscular tissues are advised. These are given in the pain zone, in pain points, in muscle gellings, and in trigger points[20] When clinical examination findings and questioning of the patient point to tendon problems, injections around the tendon are given. When ligamentous injuries are suspected, injections into ligamentous pain points or trigger points are advised. Injections into pain points can also be given at the enthesis or the periosteum.

In the general approach there are three major branches: lymphatic drainage, detoxification, and endocrine regulation. Drainage of the lymphatic system is achieved by giving subcutaneous injections of Lymphomyosot in three areas: the thorax, the liver, and the intestinal tract. The detoxification is mainly focused on the liver and the kidneys. For the liver, we start with Hepar compositum and then use Hepeel. For the kidneys, we start with Solidago compositum and then use Berberis-Homaccord. Endocrine regulation is usually focused on bringing the female cycle into balance. This is important for cycle-related complaints (e.g., migraine). We start with Hormeel injections and continue with Ovarium compositum injections.

[20] Kersschot J. *Biopuncture in General Practice*. Aartselaar: Inspiration; 2004:26–41

Follow-up of the treatment is done by checking the actual complaints of the patient and by clinical examination. If one particular antihomotoxic medication does not give the desired clinical effect, we switch to another product until resolution is reached.

*Overview of Injection Techniques*

**Subcutaneous Injections**

Subcutaneous injections are used regularly in biopuncture. According to Michel Pistor, a French physician, we mainly use antihomotoxic remedies rather than conventional medication.

***Subcutaneous Injections for Musculoskeletal Disorders***

Subcutaneous injections are given when deeper injections are impossible for technical or practical reasons (e.g., they are less painful than deep injections). Subcutaneous injections are also given for sports injuries (e.g., muscle strain) in order to use the cutaneomuscular reflexes to influence deeper layers underneath. In a later phase, deeper injections (e.g., into muscles and ligaments) can be used.

***Subcutaneous Injections for Organ Disorders***

Biopuncture stimulates and regulates local healing mechanisms by giving the medications as close as possible to the target. In some cases, however, it is impossible to inject right into the target organ, which is why we have to use injections at a distance, similar to other reflex therapies such as mesotherapy. In biopuncture, we inject biotherapeutic agents into points that lie in the zone related to the relevant organ. Thus, we administer subcutaneous injections anywhere into the reflex zone of a particular organ. Usually, these reflex zones anatomically overlie the organ involved. Reflex zones can be both on the front side and the back side of the body.

For example, Traumeel S is injected into the reflex zone of the organ we want to influence and regulate when inflammations are involved. More specific products can be used if appropriate. These products include Nux vomica-Homaccord, Hepeel, Drosera-Homaccord, Hormeel, Engystol, Echinacea compositum, and Sabal-Homaccord. When treating asthma or bronchitis, Traumeel and/or Engystol can be injected subcutaneously into the reflex zone of the bronchi. When treating gastritis, Nux vomica-Homaccord injections are administered into the gastric area, three finger-widths under the xiphoid. When treating patients with colon spasm, Spascupreel injections can be administered into the abdominal area and/or the lumbar area. When treating cystitis, Traumeel or Berberis-Homaccord is injected subcutaneously in the lower abdomen and/or lumbosacral area. In other words, subcutaneous injections are given to influence the underlying organs (segment therapy).

#### Intramuscular Injections

When caring for patients suffering from minor orthopedic complaints, biopuncturists focus largely on the muscular system. The major complaints related to myofascial syndrome are myofascial pain in the involved muscles, weakness in those muscles (less strength), and tightness in the muscles (less stretching)[21]. Muscular pain can produce both local and referred pain. *Referred pain* means that the pain is sent to a distant part of the body. This is called the pain referral zone of a trigger point. Travell and Simons have described the pain referral pattern for each muscle.

The standard treatment for myofascial pain is local intramuscular injections with Lymphomyosot and Spascupreel. If it is posttraumatic, Traumeel can be used instead of Spascupreel. Depending on the size of the area, from 1 to 10 injections can be given into the regional muscles during each session. The depth varies with the depth of the tender spots in the muscle involved. If biopuncturists want to influence different layers of a muscle, they place the needle into the deepest area and then withdraw the needle while injecting continuously.

To avoid injections into nerves, vessels, or organs, "pincer palpation" is used. The patient's muscles are taken between thumb and forefinger and the needle does not penetrate the tissues any deeper than the fingertips. Pincer palpation also makes the injection less painful.

[21] Kersschot J. Trigger points: detection and treatment. Symposium Orthopedic Medicine, Brussels, October 14, 2000

#### Tendon Injections

In acute and uncomplicated cases, the injection is made into the pain zone (at random in the area) with Traumeel and Lymphomyosot. These injections are not given *into* the tendon but rather *around* the tendon (e.g., Achilles tendon). We can also look for specific tender spots and inject these with the same cocktail. In chronic cases, Coenzyme compositum and Zeel are also used.

#### Infiltrations of Ligaments

When ligament fibers are overstretched, the sensory nerve endings within the tendon are irritated. This usually related to a traumatic event (e.g., a car accident or a sports injury as a result of a fall). Especially when no bone structures are broken, the impact of the traumatic event will be absorbed by the ligaments[22].

Traumeel is the first-choice medication; Zeel and Coenzyme compositum can also be used. Local anesthetics and hypertonic sugar water can be added. The final concentration of hypertonic sugar is usually 10% dextrose. When injecting in the middle of a ligament (e.g., nuchal ligament), we inject as close to the ligament as possible, at several spots. When injecting near the attachment of the ligament to the bone (enthesopathy), we point the needle until bone contact and then withdraw the needle 1 mm and inject. By injecting Traumeel, we evoke an inflammatory response that starts the healing cascade.[23] The hypertonic dextrose added to the Traumeel S is important for stimulating fibroblastic activity in the area (proliferant effect).

[22] Kersschot J. Injections of hypertonic solutions for ligament injuries. Congress of Sports Medicine, Val d'Isère (France), April, 2007

[23] Van Brandt B. *Inflammation Means Healing.* 3rd ed. Aartselaar: Inspiration Publishing; 2004

*Practical Considerations*

Biopuncture can only be practiced by doctors who are licensed and insured to give these injections. The standards and laws regarding injections for medical purposes and the registration of these products may differ from country to country (and from state to state), and therapists are always advised to check the local regulations on these techniques. All injections are given only after a medical diagnosis, and the injections themselves are administered according to the laws of orthodox medicine, especially concerning safety and sterility. In biopuncture training, licensed therapists are informed about the strategies, contraindications, and pitfalls in this field so that they may achieve the best results and to prevent complications.

*Contraindications*

Biotherapeutic injections are not given when patients do not wish to undergo injection (e.g., in cases of needle phobia) or when patients have a very bad physical or mental condition (e.g., unstable heart rate, high blood pressure, poor respiratory system, or abnormal sugar metabolism). We do not inject in areas where we do not feel safe about the nature of the tissues and their structure, such as neoplastic lesions, deformations (e.g., after an operation or an accident), and fractures. We do not inject patients with bleeding diathesis (especially when giving deep injections) or patients who have an allergy to one of the ingredients.

*Technical Complications*

Technical complications include allergic reactions, skin rash, hematomas, swelling of the injected area, and local stiffness. Infections after injection are avoided by using sterile injection techniques.

Damage to organs can occur after accidental injections into the pleura, kidney, liver, and intestinal tract. These complications can be prevented by performing a competent clinical examination, injecting with respect to anatomical limits, using radiological control before or during the injection, taking the thinnest needle possible, and using safe injection techniques (i.e., not injecting close to organs such as the lungs or kidneys).

Accidental damage caused by the needle to nerves or small blood vessels cannot always be prevented. Serious complications caused by inadvertent injections into important blood vessels or the epidural space are very rare among biopuncturists because unsafe injection techniques are avoided at all costs in the biopuncture course.

*Conclusions*

Biopuncture is not a new therapy; it is a unique combination of several strategies with very interesting therapeutic possibilities. Because more physicians worldwide are practicing biopuncture, more and more patients can benefit from the effects of these biotherapeutic injections. This will lead to more exposure of this technique and to more scientific investigation in the field of antihomotoxic injections.

## Tried and Tested Indications and Special Treatment Strategies

### Respiratory Diseases

The treatment of everyday colds with preparations from antihomotoxic medicine is highly recommended. A whole range of tried and tested combination preparations are available. The basic medication is undoubtedly the preparation Gripp-Heel, the antiviral efficacy of which has been confirmed in laboratory tests. The preparations Euphorbium compositum, Engystol, and Lymphomyosot should also be considered as acute medications. Lymphomyosot has proved particularly valuable in treating children susceptible to infections.

### *Oral Autologous Blood Therapy*

Autologous blood therapy should be given for repeated viral diseases. This is of course not popular with children, so oral autologous blood therapy may be a welcome alternative. The following strategy has proved successful in my practice:

Take a 30 mL vial of Lymphomyosot and divide it into equal 2.5 mL parts in 12 small 5 mL dropper bottles. Place one drop of autologous blood in the first bottle and shake vigorously 10 times. From the resultant homeopathic dilution take one drop, place it in the second bottle and shake again. Take a drop from the second bottle and add to the third bottle, then shake. Continue in this way with the serial dilution until you reach the twelfth bottle. For the treatment of susceptibility to infection, start with the sixth bottle in the serial dilution and prescribe the patient 10 drops 3× a day until the bottle is used up. This should be followed by the seventh bottle, also with 10 drops 3× a day. From the eighth bottle onward, reduce the dose to 3×5 drops; from the tenth bottle onwards, to 1×5 drops a day; and from the twelfth bottle prescribe 5 drops 1–3 times a week.

### *Progressive Auto-sanguis Therapy in Recurring Viral Infections*

1. Gripp Heel
2. Lymphomyosot
3. Mucosa compositum
4. Nosode (Grippe Nosode Injeel, Herpes simplex Nosode Injeel) if available

Twice a week, a total of 5–10 times.

### *Progressive Auto-sanguis Therapy in Recurring Bacterial Infections*

1. Traumeel
2. Lymphomyosot
3. Mucosa compositum
4. Echinacea compositum

Twice a week, a total of 10 times.

The last step in this cure can be infiltrated into the Shu point of the lung (BL-13) on both sides, which greatly improves efficacy.

Large-scale autologous blood therapies are recommended in severe cases of disease. Hematogenic oxygenation therapy (HOT), treatment with UV-irradiated autologous blood (UVB), and ozone autologous blood therapy have proved extremely successful here. It is recommended that infusion therapies be given with these autologous blood therapies:

Physiological saline or balanced electrolyte solution 250 mL
Vitamin C 3–7.5 g
+ 1 ampule each of Gripp Heel, Engystol, Lymphomyosot,
Mucosa compositum and Echinacea compositum

Twice a week, a total of 10 times.

Chronic sinusitis is very often the cause of chronic respiratory diseases, especially asthma or spastic bronchitis in infants. Euphorbium compositum has proved useful here, but other therapies can also be recommended. It is less well known that the ampule preparations in homotoxicology can be inhaled, even though that is not an official use. In practice we had very good results with a mixture of Traumeel, Engystol, Mucosa compositum, Euphorbium compositum, and Lymphomyosot for inhalation using a nebulizer. Spascupreel has also proved its worth in asthma.

### *Progressive Auto-sanguis Therapy in Chronic Sinusitis*

1. Euphorbium compositum
2. Lymphomyosot
3. Mucosa compositum
4. Sinusitis Nosode Injeel

Twice a week, a total of 5–10 times.

Infiltration of the last step into the point LI-10 is unpleasant for the patient but is extremely effective.

### Autoimmune Diseases

The adjuvant treatment of autoimmune diseases starts with excretion and detoxification therapy. Under no circumstances should an immunostimulant treatment, for example with autologous blood, nosodes or other immunostimulant homeopathic medications (such as Echinacea), be given at the start of therapy. There would be a risk of an acute exacerbation of the disease. First of all the patient's homotoxin load should be reduced as far as possible. This should be followed by targeted strengthening of the diseased organ systems; the homotoxicological composita preparations play an essential role here. Only after the symptoms have improved can the incorrectly programmed immune system be modulated with the greatest care and circumspection.

*General Excretion Infusion*

Electrolyte solution, 250–500 mL
+ 1 ampule each of Lymphomyosot, Hepar compositum, Solidago compositum and Nux vomica-Homaccord

This infusion is given once to twice a week for 5–10 weeks.

With blocked regulation, from the second infusion onwards the 10 ampules in the multipack of citric-acid-cycle catalysts are added and this is repeated 2 or 3 times at weekly intervals. The additional intramuscular injection of Coenzyme compositum and Ubichinon compositum is also promising in these cases.

The cause of autoimmune diseases can often be found in the patient's chronic infections or chronic exposure to toxins. Herpesviruses, influenza viruses, coxsackieviruses or *Candida*, *Borrelia*, streptococci, and staphylococci are often found in this connection. Exposure to pesticides and herbicides is also found. Drugs taken in the past, such as antibiotics, immunosuppressants, and symptom-blocking allopathic medications are repeatedly suspected of being involved in triggering autoimmune diseases.

These diseases can ultimately be cured only if the antihomotoxic therapy is given consistently; this means that nosodes and immunomodulating measures ultimately also play an important role.

*Progressive Auto-sanguis Therapy in Allergies*

1. Histamin-Injeel + Urtica-Injeel
2. Lymphomyosot
3. Mucosa compositum and/or Cutis compositum
4. Engystol or the appropriate nosode according to the case history

One to two treatments a week, starting about 4 weeks before the season, a total of 10 times. Supporting therapy is given with Luffa compositum tablets and nose drops. Apis-Homaccord and Formica rufa Injeel have proved their worth in allergies.

### Diseases of the Locomotor System

Antihomotoxic medicine and in particular injection therapy play a central role in this organ system. Experience shows that injections close to the diseased joint, connective tissue, or muscles produce the best results. Injections at defined acupuncture points produce a faster onset of action and prolong the success of the treatment. Important injection sites are the alarm points, the Shu points and, in joint diseases, the classic distal points LI-4, ST-36 and GB-34. Injections into microacupuncture systems such as the ear points, the scalp points according to Yamamoto, and the mouth points according to Gleditsch often produce amazing immediate effects.

Rheumatic diseases are often accompanied by a microcirculation disorder and hyperacidity of the connective tissue. For this reason, infusion therapy is often given at the start of treatment.

*Infusion in Degenerative Joint Diseases*

Balanced electrolyte solution 250 mL
Procaine 1 % 10 mL
Zeel 2 ampules
Traumeel 1 ampule
Lymphomyosot 1 ampule
Circulo Injeel 1 ampule
Cartilago suis 1 ampule
Coenzyme compositum 1 ampule

Once to twice a week, a total of 10 times, then once a month.

*Infusion in Inflammatory Joint Diseases*

Balanced electrolyte solution 250 mL
Sodium hydrogen carbonate 8.4% 20–40 mL
Procaine 1% 10 mL
Traumeel 2 ampules
Lymphomyosot 2 ampules
Mucosa compositum 1 ampule
Echinacea compositum 1 ampule
Coenzyme compositum 1 ampule

*Progressive Auto-Sanguis Therapy in Osteoarthritis*

1. Zeel
2. Lymphomyosot
3. Cartilago suis
4. Coenzyme compositum

*Progressive Auto-sanguis Therapy in Arthritis*

1. Traumeel
2. Lymphomyosot
3. Mucosa compositum
4. Echinacea compositum and Coenzyme compositum

## Vascular Diseases

Circulatory disorders are very good indications for antihomotoxic therapy. The preparation Vertigoheel has become the standard treatment for dizziness in Germany. Its efficacy has been confirmed in numerous scientific investigations. What is particularly important in practical terms is promotion of the microcirculation by Vertigoheel. Experience shows that a higher dose initially has to be recommended and that, in acute cases, injection therapy is the stronger treatment option. The major autologous blood therapies with ozone and oxygen that supplement antihomotoxic therapy very well have assumed great importance in practice.

*Infusion Therapy in Vertigo*

HOT or ozone therapy
+ balanced electrolyte solution 250 mL
Vertigoheel 2–4 ampules
Circulo-Injeel 1 ampule
Cerebrum compositum 1 ampule
Traumeel or Discus comp. 1 ampule
*Oral supportive therapy:*
Vertigoheel 3 × 2–3 tablets or 3 × 10–30 drops
Homeosiniatry with Vertigoheel as auriculotherapy in both ears in the projection tract region of the vestibular system and the cervical spine.

*Infusion in Arterial Circulation Disorders*

Ozone therapy
+ electrolyte solution 250–500 mL
Procaine 1% 10 mL
Circulo-Injeel 2–3 ampules
Vertigoheel 1 ampule
Coenzyme compositum 1 ampule
Arteria suis-Injeel 1 ampule

Once to twice a week, a total of 10 times, then once a month.

*Infusion Therapy in Venous Circulation Disorders*

HOT
+ electrolyte solution 250 mL
Calcium 10% 10 mL
Aesculus Injeel 1–2 ampules
Lymphomyosot 2 ampules
Tonico-Injeel 1 ampule

## Skin Diseases

The main treatments for skin diseases are the excretion therapies and the minor autologous blood therapies.

*Excretion Infusion in Skin Diseases*

Balanced electrolyte solution 250 mL
Zinc 10 mL ampule
Cutis compositum 1 ampule
Lymphomyosot 1 ampule
Engystol 1 ampule
Graphites-Homaccord 1 ampule

10 infusions within 5–10 weeks.

*Progressive Auto-sanguis Therapy in Acne*

1. Traumeel
2. Lymphomyosot
3. Cutis compositum
4. Echinacea compositum

*Progressive Auto-sanguis Therapy in Eczema*

1. Graphites-Homaccord
2. Lymphomyosot
3. Cutis compositum
4. Engystol or Psorinum Injeel

### Heart Diseases

In heart diseases the injection of the ampule preparations into the Shu point of the heart (BL-15) has proved useful:

- Cor suis compositum in degenerative diseases of the myocardium
- Cactus compositum in stenocardic heart complaints
- Cralonin in senile heart and cardiac arrhythmia, and here also:
- Strophantus compositum

*Infusion Therapy in Senile Heart*

Manifest uncompensated heart failure is not an indication for infusion therapy; the following infusion is recommended after compensation:

Balanced electrolyte solution 250 mL
Cralonin 1–2 ampules
Cor suis compositum 1 ampule
Strophantus compositum 1 ampule
Cactus compositum 1 ampule
In addition 1 ampule each of Coenzyme compositum and Ubichinon compositum IM.

Twice to 3 times a week initially, less frequent doses after 1–2 weeks.

*Infusion Therapy in Stenocardia*

Balanced electrolyte solution
Magnesium 10 mL 1 ampule
Cactus compositum 1–2 ampules
Strophantus compositum 1 ampule
Cralonin 1 ampule
Circulo-Injeel 1 ampule

Once to twice a week, a total of 5–10 times.

### Liver Diseases

Acutely inflammatory and chronically degenerative liver diseases can be treated very effectively with antihomotoxic therapy. Infusion therapies in conjunction with autologous blood therapies have proved extremely effective at the start of treatment. In addition to the liver and gallbladder meridian, homeosiniatric treatment is given mainly at the Shu point of the liver (BL-18).

*Infusion Therapy in Hepatitis*

Ozone therapy or HOT
+ balanced electrolyte solution 250 mL
Vitamin C 5–15g
Hepeel 1 ampule
Hepar compositum 1 ampule
Engystol 1 ampule
Echinacea compositum 1 ampule
Lymphomyosot 1 ampule

Twice to 3 times a week, a total of 19 infusions.

### *Infusion in Cholestasis*

HOT or ozone therapy
Balanced electrolyte solution 250 mL
Hepar compositum 1 ampule
Chelidonium-Homaccord 1 ampule
Injeel Chol 1 ampule

Twice to 3 times a week, a total of 10 times.

### *Infusion for Liver Detoxification*

Balanced electrolyte solution 250 mL
Vitamin C 5–15 g
Hepeel 1–2 ampules
Hepar compositum 1 ampule
Nux vomica-Homaccord 1 ampule
Lymphomyosot 1 ampule
Heel citric acid cycle, all 10 ampules, every second time

Twice to 3 times a week, a total of 10 times.

## *Gastrointestinal Diseases*

Nausea and retching, heartburn, *Helicobacter pylori* infections, digestive disorders, food intolerances, and chronic inflammation are all very good indications for antihomotoxic therapy. Here are a few indications that have been tried and tested in practice.

### *Excretion Infusion for Stomach and Intestine*

Balanced electrolyte solution 250–500 mL
Nux vomica-Homaccord 1 ampule
Hepar compositum 1 ampule
Leptandra compositum 1 ampule
Podophyllum compositum 1 ampule
Mucosa compositum 1 ampule

Twice a week, a total of 10 times.

### *Progressive Auto-sanguis Therapy in Colitis*

1. Traumeel
2. Lymphomyosot
3. Podophyllum compositum + Mucosa compositum
4. Bacterium coli Nosode-Injeel (only later, after the condition has improved)

In food intolerance, additionally oral Gastricumeel and Okoubaka 2 × 1 tablet of each before meals.

### *Infusion in Heartburn and Hyperacidity*

Balanced electrolyte solution 250–500 mL
Sodium hydrogen carbonate 8.4% 20–40 mL
Traumeel 1–2 ampules
Lymphomyosot 1 ampule
Mucosa compositum 1 ampule

2–3 infusions a week initially, a total of 10 times, later 1 infusion a month. In the evening also administration of an oral base preparation (e. g., Sander's base powder).

### *Infusion Therapy of* ***Helicobacter pylori***

Balanced electrolyte solution 250 mL
Sodium hydrogencarbonate 8.4% 20 mL
Traumeel 1 ampule
Engystol 1 ampule
Mucosa compositum 1 ampule

1–2 infusions a week, a total of 10 times, also 3 × 2 Gastricumeel tablets orally before food.

### *Progressive Auto-sanguis Therapy of* ***Helicobacter pylori***

1. Traumeel
2. Lymphomyosot
3. Mucosa compositum
4. Echinacea compositum + Engystol

Once to twice a week, a total of 10 times. Also a base powder in the evening.

### *Urogenital Diseases*

*Progressive Auto-sanguis Therapy of Recurring Urinary Tract Infections*

1. Traumeel
2. Galium-Heel + Lymphomyosot
3. Solidago compositum and/or Cantharis compositum
4. Mucosa compositum + Echinacea compositum or Bacterium coli Nosode-Injeel

2 treatments a week, the last stage can be infiltrated into the bladder Shu point (BL-25) or suprapubically.

*Progressive Auto-sanguis Therapy in Failure to Conceive*

1. Traumeel
2. Lymphomyosot
3. Mucosa compositum
4. Ovarium compositum or suis or Testis compositum
5. In addition to these, consistent excretion therapy

1–2 treatments a week, a total of 10 times. The last stage can be infiltrated into the bladder Shu point BL-25 or the suprapubic region.

With hormonal disorders, also Hormeel orally or in ampules plus Hypophysis suis and Hypothalamus suis-Injeel, 2 mixed injections a week.

With hormonal disorders, 1 or 2 mixed injections of these suis-organ preparations a week have proved successful: Hypophysis, Hypothalamus, Glandula thyreoidea, Glandula suprarenalis, Ovarium or Testis.

*Infusion for Renal Excretion*

Electrolyte solution 250–500 mL
Solidago compositum 1 ampule
Lymphomyosot 1 ampule
Mucosa compositum 1 ampule
Coenzyme compositum 1 ampule

Twice a week, a total of 10 times.

## *Prospects*

Antihomotoxic medicine offers a wealth of different treatment options and is an important component in any holistic treatment concept. This special treatment method has its roots in homeopathy but addresses the special challenges of our modern times.

Samuel Hahnemann urged colleagues, "Copy what I do, but copy it exactly." This incitement is misunderstood by some homeopaths as a command by Hahnemann to follow his therapy religiously. However, even today one can easily divine from the works of Hahnemann the doctor who was always critically reexamining the very essence of his treatment and changing it in the light of improved knowledge.

Hahnemann could not have predicted many of the challenges of our modern age, particularly chronic diseases. New treatment methods such as antihomotoxic medicine are in tune with our times.

Our therapy lives on experience, but can also increasingly become established in scientific medicine. The research activities of the company Biologische Heilmittel Heel support the ongoing task that company founder Hans-Heinrich Reckeweg set himself, that of merging homeopathy with mainstream medicine.

The most important foundation of our therapeutic systems, however, is our own experience and the international exchange of views with colleagues. Ongoing and further training activities worldwide provide welcome opportunities for doing this. Those running curricular training events and study courses learn what students have accepted and how successful treatments have proved for them in practice.

# Index

## A

## B



## C



## D